Diet and Crafts in Towns

The evidence of animal remains from
the Roman to the Post-Medieval periods

edited by

Dale Serjeantson and T. Waldron

BAR British Series 199
1989

B.A.R.

5, Centremead, Osney Mead, Oxford OX2 0DQ, England.

GENERAL EDITORS

A.R. Hands, B.Sc., M.A., D.Phil.
D.R. Walker, M.A.

BAR 199, 1989: 'Diet and Crafts in Towns'

© The Individual Authors, 1989

The authors' moral rights under the 1988 UK Copyright,
Designs and Patents Act are hereby expressly asserted.

All rights reserved. No part of this work may be copied, reproduced, stored,
sold, distributed, scanned, saved in any form of digital format or transmitted
in any form digitally, without the written permission of the Publisher.

ISBN 9780860545989 paperback
ISBN 9781407318196 e-book
DOI https://doi.org/10.30861/9780860545989
A catalogue record for this book is available from the British Library
This book is available at www.barpublishing.com

CONTENTS

	Page
INTRODUCTION	1
WHAT SHALL WE HAVE FOR DINNER? FOOD REMAINS FROM URBAN SITES by Terry O'Connor	13
THE PROVISION OF FOWLS AND FISH FOR TOWNS by Jennie Coy	25
SOCIAL DIFFERENTIATION FROM ANIMAL BONE STUDIES by F. Gerard Ijzereef	41
THE EFFECTS OF HUMANISATION ON HEALTH: THE EVIDENCE FROM SKELETAL REMAINS by Tony Waldron	55
URBAN RURAL VARIATIONS IN THE BUTCHERING OF CATTLE IN ROMANO-BRITISH HAMPSHIRE by Mark Maltby	75
BONE, ANTLER AND HORN INDUSTRIES IN THE URBAN CONTEXT by Arthur MacGregor	107
ANIMAL REMAINS AND THE TANNING TRADE by Dale Serjeantson	129
THE USE OF ANIMAL BONES AS BUILDING MATERIAL IN POST-MEDIEVAL BRITAIN by Philip L. Armitage	147
BONE ANALYSIS AND URBAN ECONOMY: EXAMPLES OF SELECTIVITY AND A CASE FOR COMPARISON by Bruce Levitan	161
DECIDING PRIORITIES WITH URBAN BONES: YORK AS A CASE STUDY by Terry O'Connor	189
GAZETTEER OF SITES WITH ANIMAL BONES USED AS BUILDING MATERIAL by Philip L. Armitage	201

ADDRESSES OF CONTRIBUTORS

Dr. Philip L. Armitage
1972 Roseate Lane
Sanibel Island
FLORIDA 33957
U.S.A.

Jennie Coy
Faunal Remains Unit
Department of Archaeology
University of Southampton
Southampton SO9 5NH

Dr. F. Gerard Ijzereef
Zooarchaeologist of the Dutch State Service for Archaeology
Rijkdienst voor het Oudheidkundig Bodemonderzoek
Kerkstraat 1
3811 CV Amersfoor
Netherlands

Bruce Levitan
University Museum
Parks Road
Oxford OX1 3PW

Dr. Arthur MacGregor
Department of Antiquities
Ashmolean Museum
Oxford OX1 2PH

Mark Maltby
Faunal Remains Unit
Department of Archaeology
University of Southampton
Southampton SO9 5NH

Dr. Terry P. O'Connor
Environmental Archaeology Unit
Department of Biology
University of York
York YO1 5DD

Dale Serjeantson
Birkbeck College
Centre for Extra-Mural Studies
University of London
26 Russell Square
London WC1B 5DQ

Dr. Tony Waldron
Institute of Archaeology
31-34 Gordon Square
London WC1H 0PV

INTRODUCTION

Dale Serjeantson

AIMS

During the past twenty years the investigation of archaeological animal bones has been common, yet comparatively little, apart from a survey by Armitage (1982), has concerned the methods of study of bones from urban sites.

Consequently excavators of such sites have had few opportunities to learn how animal remains can contribute to the understanding of the archaeology of the town or site under investigation. In this book we have aimed to go some way to filling this gap. Its origin was a conference held at the Department of Extra-Mural Studies of the University of London in December 1986. Here students had worked in class on animal remains from a variety of types of sites, and it had become clear that most of the work on theoretical aspect of analysis of animal bones has been directed at collections from prehistoric sites in a rural setting.

The conference organisers therefore sought contributions which would cover a wide range of the problems relating to the aims and methods of the study of animal remains from urban sites. The archaeological remains from towns have two particular characteristics: there is specialisation in crafts and industries, and excavations are usually of multi-period deposits. Though most of our contributors are from Britain and their examples are mainly from Europe, the material and methods discussed should be relevant to towns everywhere.

In the main bones were most often thrown away after food preparation and consumption, and from these much can be learned about the diet and also the social composition of the urban population. However, especially in towns, bones were also discarded as by-products of craft processes. In some circumstances the most important contribution which the bone analyst can make is to assist in the interpretation of the context which is being studied and the activities which led to the formation of the excavated deposit. Animal remains from urban excavations demand different methods of recovery and analysis from those in a rural context. There are papers on all these aspects. Additionally, increasing numbers of early cemeteries are threatened by urban re-development, and we thought it appropriate also to include a paper on the information which the excavation and study of human skeletons could provide.

THE STUDY OF DIET AND SOCIAL STATUS

A number of contributors have written about the relationship between animal bones and diet in towns. O'Connor (pp. 13-23) shows that from the Middle Ages onward most of the meat consumed was from cattle and sheep. The changes charted in the relative numbers of the two species at Lincoln and York show how sheep increased at the expense of cattle, and it will be important to see if this trend is found throughout the country or only in the North. In York and Lincoln the sample was the whole town, and this unit of study is appropriate where samples are large enough and from a range of contexts within the town. In his paper on two areas of Amsterdam Ijzereef (pp. 41-54) identifies changes at a household level; he was able to distinguish not only Jewish and non-Jewish households, but also their relative wealth.

Two contributors consider how food from animal sources was brought in to towns. O'Connor makes the useful distinction between the cattle and sheep, which were raised outside the towns and driven in to the markets, and pigs, which at least in the Middle Ages were often raised by individual households in the town itself. Herds and flocks were driven often very long distances to town markets. From the seventeenth century onwards cattle raised as far as the West of Scotland and Ireland were driven to southern Britain. The implications of this trade for the study of cattle and sheep bones has not been explored in this book, but has been investigated elsewhere by Armitage (1980).

Coy points out that, unlike cattle and sheep, domestic fowls were kept in towns, and the numbers of chicken bones are often high among urban food remains. Other domestic birds are less often found; in the Middle Ages those associated with high status households such as peafowl (Pavo cristatus), swan (the Mute Swan, Cygnus olor) and turkey (Meleagris gallopavo) are as likely to be found in castles and monasteries as in towns.

Except for those who lived by the coast, fish only become important in the European diet from the early Middle Ages onwards. The herring fisheries in the Baltic and the North Sea developed from that time, and the Atlantic and offshore fishing for cod and related species became increasingly important in the economy of the north European countries with the development of seagoing boats. It is difficult to see how the restrictions on meat eating which originated with St Benedict in the sixth century could have been observed if herring and other fish has not been available to take the place of the 'flesh of quadrupeds'. Sea fish were traded over long distances: the Southampton port records, discussed here by Coy are a good illustration of this.

Though herring and cod were the main species traded and eaten, Coy shows that the range of species landed at Southampton and traded through her ports was very great. In these days of rapid transport methods and refrigeration it is easy to underestimate the range of fish which were preserved by such means as salting, salting and drying ("stockfish"), and pickling. When bones of cod, ling and herring are recovered in excavation in towns many miles from the coast they are readily interpreted as being from preserved fish, but it is also commonplace (provided sieving is carried out, cf pp. 25-27) for bones of sea fish such as the flatfish to be found, and documentary evidence suggests that these too are likely to be from preserved rather than fresh fish (Cutting 1955). From Anglo-Saxon times the systematic capture of eels is recorded in the charters, and some examples from archaeological contexts suggest that they continued to be taken in quantity. In medieval Southampton eel bones were second most numerous after herring bones, and the same relative abundance has been found in a fourteenth century context at St Albans (Serjeantson forthcoming). The importance of fish in the diet is shown in Amsterdam, where fish bones are 10% by weight (Ijzereef, table 1) in the seventeenth and eighteenth centuries pits. As O'Connor indicates, wild foods other than fish are rare. Venison was not a commodity bought and sold but, as Rackham's documentary study (1986) shows, was part of the gift exchange network between families of high rank. This is confirmed by finds of deer bones: they are found in greatest numbers not in towns but in castles, such as Barnard Castle (Jones & Sly, N.D.), where they were almost as numerous as cattle bones, and Abbeys. Dietary deficiency, which is often associated with urbanisation, cannot be detected from food remains alone, as those which survive are only a partial record of what was consumed; but, as Waldron (pp. 55-73) makes clear, may in some cases be detected in the human skeleton.

Recognition of remains of bones from food consumption

How may food remains be distinguished from deposits from butchery or other activities, or from a mixture of origins? Four criteria are suggested here by which deposits of food remains may be recognised. It is in fact rare for all to be fulfilled, because groups of bones from town sites are rarely from truly primary contexts, but some at least should. The four are:

1. The context is a residential building.

2. The bones are exclusively or mainly of food animals.

3. There is evidence of butchery.

4. The parts of the skeleton which have the most meat predominate.

To consider each criterion in detail:

1. The Amsterdam cess-pits studied by Ijzereef are a good example of bones in a domestic context. However, many buildings which were not domestic households, such as abbeys, were also lived in, and often provide good examples of assemblages of food remains. For instance the assemblage from Whitefriars, Coventry, (Holmes 1981) which is discussed below, was a school. Where rubbish was disposed of on the periphery of the town, dumps can often contain material from a single source, either domestic or craft rubbish.

2. A high proportion of fish and domestic fowl bones is likely to indicate food remains. Conversely bones of dogs, cats, rats and mice do not. Horse bones too in most contexts are unlikely to be food remains, though they may be, particularly on Roman sites. In this case evidence of butchery may confirm whether or not they were eaten.

3. Disarticulation of the carcase and food consumption leaves some individual bones unmarked, but there will be a proportion in the group which has been chopped. Whole bones in articulation are likely to be from animals which have died and were not eaten.

4. A high proportion of the main meat bearing bones is the criterion used often to distinguish waste from meat consumption from butchery waste. With the proviso that some are associated with crafts, in general the head and feet have least meat and the rest of the skeleton has most.

However, there are some factors which may complicate the analysis.

Firstly, the interpretation may be affected where deposits were not sieved. The effect this has on estimating the contribution to the diet of fish has already been discussed. But it can also seriously distort the interpretation of sheep remains. The scarcity of teeth and phalanges is a criterion that bones are mainly from food consumption. However, this lack in the body parts present of sheep can be a consequence of poor retrieval, usually when deposits are not sieved (Payne 1972).

Secondly, where bones have been heavily gnawed, the composition of the assemblage is not uniformly affected. Dogs consume the more fragile bones preferentially to the more dense shafts and articulations. Ribs, vertebrae, and the femur and proximal tibia are more likely to be destroyed than the distal humerus and distal tibia. Dogs are least likely to destroy mandibles and teeth (Brain 1976, Binford and Bertram 1977). The effect of dogs on an assemblage is to depress the relative numbers of skeletal elements from regions which were originally choice cuts, such as the vertebrae, and leave a spuriously high proportion of jaws and teeth.

A good example of a deposit of food remains is the assemblage from the Resonance Chamber of Whitefriars School, Coventry. Possibly as few as a dozen bones from a total of 1880 are from species unlikely to have been eaten, including mouse and cat (Holmes, table 1), whereas rabbit and bird bones are numerous. Skull and foot bones of the main food animals are scarce and 93.7% of sheep bones are from the main meat bearing bones. The material is not in a primary position, but it is relatively unmixed. There is no record of sieving taking place, which is likely to account for the small number of fish remains.

THE DEBRIS FROM CRAFTS

The earliest record of a specialised archaeological deposit of bones must be the collection of "a hundred ox skulls" reputedly found in the early fourteenth century when old St Pauls was rebuilt (Clark 1980). It is interesting to speculate whether the builders had come upon a ritual deposit or rubbish in a Roman ditch (cf Maltby p. 75).

Butchery

Butchery was often more specialised in towns than it was in the country. Maltby (pp. 75-106) shows that in the centuries of Roman control different butchery techniques were employed in the town and in the countryside: town butchers employed crude chops with heavy axes to split the carcass, whereas in the villages the ancient tradition using knives was more often used to disjoint the animals for consumption. Butchery continued to be done with heavy metal cleavers in the early Middle Ages. The main innovation after that is in the later Middle Ages when butchers began to split the carcase down the midline (Grant 1987, Maltby 1979).

Identifying the waste from primary butchery is rarely as straightforward as in the example from Exeter (Maltby 1979, 11), where a dump in one of the defensive ditches of the legionary fortress consisted predominantly (87%) of cattle skulls minus the horn cores, and metapodials minus the foot bones. More often the distinction is less clear; this is not just because discarded waste often contains material from more than one source, but because all parts of the carcase (except the horn and horn core) may be cooked and eaten, and so may be found in the waste from food consumption.

The trades which made use of animal products in the City of London in the eighteenth century (table 1) shows that a greater variety dealt in animal products than in foodstuffs. The list has been compiled from a sample of 14% of the entries in the Universal British Directory (1791). Craft with which characteristic bone remains are associated are horn-working, bone and antler working, and tanning. Extraction of oil and grease from bones may also belong in this category.

Antler, bone and horn-working

Remains from antler working and bone working are not very common but are easily recognised, by the characteristic marks of sawing. Some offcuts from bead making and button making have been illustrated (MacGregor this vol., figure 3; 1985, figure 58).

Much more common and also easily recognised are collections of horn cores. As MacGregor (pp. 117-119) points out, the obvious interpretation is that these are waste from hornworking, discarded after the horn itself had been taken off the core. This is thought to be the origin of a large group from Exe Bridge, Exeter, (Levitan p. 164). Positive evidence that the horn was removed from the core before it was buried is provided by cut marks round the base of the horn core or a surface sawn through, (eg. Schmid 1972, Armitage & Clutton-Brock 1976). However, such marks are rare, as separation of the horn from the core was usually achieved by rotting or soaking, which leaves no trace (MacGregor p. 117). It has also been suggested that tanners were responsible for discarding horns, as discussed by Serjeantson (pp. 136-139). Horn cores and other parts of the skeleton were used as building material in the eighteenth century, as Armitage shows in his discussion (pp. 147-160) and gazetteer (pp. 209-220).

Leather Working

Collections of metapodials and/or phalanges are often likely to have been discarded from skins; the best example of a find at a known tannery site is the sheep foot bones from Walmgate, York, but several others have been recorded (Serjeantson pp. 136-139). Where foot bones are found with no evidence for the rest of the skeleton the possibility that they are from a skin always needs to be considered.

Finally the possibility that a specialised type of butchery found on some Roman sites represents not food preparation but extraction of bone grease should be considered here. Maltby (p. 90) describes how at Winchester and Tower Street, Cirencester, the four main limb bones of cattle and horses, the humerus, radius, femur and tibia, were chopped axially and split. This butchery technique, also seen at Zwammerdam, Holland (van Mensch 1974), Balkern Street, Colchester, 24-30 Tanners Row, York, (O'Connor p. 195), and Great Chesterford (Serjeantson, in press). The degree of fragmentation is greater than that needed to extract the marrow. Though confirmatory evidence has not been found, it seems likely that the bones were chopped to facilitate the extraction of grease. The technique appears to be restricted to sites of the Roman period and to certain locations only, and Maltby suggests convincingly that it is a technique of the Roman army and the larger towns. If this chopping was prior to extracting grease from the bones, then the debris from the crafts which extracted animal fats, oils, and grease, such as tallow chandlers and soapmakers (table 1) has yet

```
FOOD                              SKINS AND FURS

   Butcher                           Skinner
   Drysalter                         Furrier
   Portable soup-maker               Tanner
   Cheesemonger                      Leather dresser
   Butter factor                     Vellum binder
                                     Leatherseller
OIL AND GREASE
                                  HAIR AND FEATHERS
   Tallow chandler
   Spermaceti refiner                Horsehair manufactury
                                     Hair merchant
HORN AND BONES                       Eiderdown warehouse
                                     Wholesale quill dealer
   Horn manufacture
   Whalebone cutter
```

Table 1. Trades which used and sold animal products. The list is from the London Directory of the Universal British Directory of 1791. London is not typical as the number of tradesmen involved in preparing and selling food were outnumbered by those making and selling craft and industrial products.

to be identified in later centuries.

Some species which are otherwise rare among the remains from towns may be represented among waste products from crafts. Though cattle and sheep were the most important food animals in Roman and Medieval towns, horses were important for transport of people and goods, the latter especially after the invention of the horse collar in the twelfth century which made horses more efficient than oxen for pulling carts (Langdon 1984), and horse bones are often found on urban sites. The butchery referred to above shows that in Roman times horse bones were treated in a similar fashion to those of cattle, which suggests that they were eaten. In the eighth century the eating of horseflesh was forbidden to Christians by Pope Gregory III, and from that time it has been culturally unacceptable in Britain and elsewhere in Europe (Barclay 1980). Bones are often found unbutchered and in articulation. Though horseflesh was apparently not eaten, the hide and the hair were used (table 1).

In the towns where goat horn cores have been found, such as Exeter (Exe Bridge, Levitan p. 163) Skeldergate, York, (O'Connor 1984), and Middle Saxon Ipswich (Jones & Serjeantson N.D.), the number of goats represented by the horn cores is higher than the number which the other parts of the skeleton suggest, and it is apparent that goat horns certainly, and goat skins probably, are being brought into the towns for processing.

A further example of a species whose remains in towns are found as craft debris but rarely as food is red deer. Antler working as MacGregor shows was based on the acquisition of shed antlers, systematically collected from forests and parks (pp. 107-110, 113).

METHODS OF ANALYSIS

One of the problems which faces excavators of town sites is how cost-effective it is to study the animal bones. So long as soil conditions are reasonable, excavations in towns produce large quantities of bones, and the resources which would allow <u>all</u> the bone to be recorded and studied have never been available. Nor is it necessarily desirable that they should be. O'Connor was therefore invited to contribute a paper on 'Priorities in urban bone studies' based on the experience of the animal bone work of the Environmental Archaeology Unit at York.

The central theme of the paper is that some form of sampling of the material is necessary. Provided survival is good and retrieval adequate, he considers that the first criterion for selection is contexts with primary material, that is closely dated deposits and deposits with material which can be related to a single activity. Groups of bones from towns have rarely been found in what would be considered a strictly primary context as defined by Schiffer (1987, 98); that is 'discarded at their location of use': contexts with bones from a single source are in this case most informative samples.

The contexts least worth study are those which contain a high proportion of residual material. Though pottery and other artefacts from such contexts may be of interest in themselves, the occasions where that is true for animal bones are so rare as to be considered non-existent for all practical purposes.

Another method of sampling to make study more cost effective is to restrict the study to certain anatomical elements. This is often done, implicitly or explicitly, and is advocated in a recent general book on archaeozoology (Davis 1986). It works well if the desired information is the identification, age, sex and type of the animal, and is an adequate basis for the study of relative numbers of the species. It is less suitable where the aim of the study is the nature and interpretation of the context, for in this case the character of the fragments which are less diagnostic zoologically may be significant to the archaeologal interpretation.

METHOD OF STUDY

A provocative suggestion made by O'Connor is that an individual record of each bone is unnecessary, and can be abandoned in favour of a summary record for each context. This is a radical departure from common practice, and the large assemblages from towns are clearly a suitable sample on which it can be tried out.

One analytical tool used widely in bone studies is the calculation of Minimum Number of Individuals. As O'Connor points out, this is nearly always inappropriate for the interpretation of bones from towns, where we are not dealing with flocks and herds or whole animals but with butchered cuts of meat, so that a more appropriate unit of measurement is the joint of meat. For example, Lyman (1987) re-analysed material from nineteenth century Sacramento in terms of a minimum number of beef cuts (MNBC) and concluded that there was a positive relationship between choice of cut and social standing.

Weighing bones to establish the relative proportions of the main species has been used since 1956 (Uerpmann 1973). It has been used effectively as a method of comparison by Ijzereef (pp. 41-54) in his work in Amsterdam. Having both counted and weighed a sample of the bones studied, he based the calculations he used in his study on weight of identified bones alone. As bone survival varies according to burial conditions, bones with the same dimensions have different weights according to whether they have been in, for instance, acid soil where bone mineral is lost or alkaline soil where it is preserved. The method can therefore only be used to make comparisons within sites or between sites where soil conditions are uniform.

Recovery

It is now routine to sieve in research excavations, but the decision to sieve a context is not taken lightly by excavators working in rescue conditions. However, there are contexts where sieving must be done if the bone evidence is to be correctly interpreted. There is no room for discussion on this point where the fish bones are to be studied (Coy pp. 25-27). How body part analysis of sheep bones is distorted when deposits are not sieved is noted above.

A comparison between the sheep foot bones from Walmgate, York, where one pit (context 1097) was sieved (O'Connor 1984), and those from Knapp Drewett, Kingston, which was not (Serjeantson et al. 1986), shows what information is lost when sieving is not done. In table 2 the numbers of metapodials and phalanges from the two sites are compared. When allowance is made for the different numbers of bones in the foot, it can be seen that

Walmgate, York			Knapp Drewett, Kingston		
	(n)	(%)		(n)	(%)
Metacarpal prox	47.5	18	Metacarpal diaphysis	29	100
Metacarpal dist	50.5	19.2			
Metatarsal prox	17	6.4	Metatarsal diaphysis	27.5	94.8
Metatarsal dist	30	11.4			
			Metapodial epiphysis	21.4	73.7
			Phalanx 1 epiphysis	15	51.7
Phalanx 1	263.5	100	Phalanx 1 diaphysis	22.75	78.4
			Phalanx 2 epiphysis	8.6	29.7
Phalanx 2	252	95.9	Phalanx 2 diaphysis	15.25	52.6
Phalanx 3	233	88.4	Phalanx 3	16.6	52.6

(n) Number of bones after taking into account the expected number in the skeleton.

(%) The most frequent bone is taken as 100% and the rest expressed as a percentage of that figure.

Table 2. Numbers and percentages of foot bones at Walmgate, York, and Knapp Drewett, Kingston, compared, to show the effect of sieving. The Walmgate assemblage gives the true picture of what bones were discarded; the percentages of the Kingston bones reflect the relative size of the bones. (Data from O'Connor 1984 and Serjeantson et al 1986.)

at Walmgate (column 1) phalanges are most numerous, with metapodials less so. From this O'Connor was able to deduce that the skins were brought to the tannery with the foot bones on, but only a proportion also had the metapodials attached. The relative proportions of calf foot bones from Kingston (column 2) are different. Here the epiphyses are unfused, so the number of bone element numbers is higher. The metapodial shafts are most common, followed by the shaft of the first phalanx, and the epiphyses of the metapodials. Least numerous were the epiphyses of the phalanges. The correlation is with the size of the bone element, and the likely interpretation is that the smaller bones were not seen in the gound and were not recovered. Here the evidence of the bones themselves could not be used to deduce whether all the toes had originally been attached. The methods of bulk sieving now being used at York which permit large quantities of even heavy soils to be sieved rapidly should make it easier for the decision to be taken to sieve contexts where the bones are likely to be important.

CONCLUSION

The quantity of bones from town sites will continue to be prodigious and will continue to outstrip the resources available even to catalogue them. Throughout this book contributors have argued that it is important to develop strategies which will extract the maximum information from the bones. The book highlights some of the more important aspects of town life which the study of animal bones can illuminate.

Acknowledgements

We would like to thank Tony Legge for arranging for the conference on which this book is based to be held at the Department of Extra-Mural Studies (now the Birkbeck College Centre for Extra-Mural Studies, University of London). We are also grateful to Harvey Sheldon for reading and commenting on the papers.

REFERENCES

ARMITAGE P.L. (1980) A preliminary description of British cattle from the late twelfth to the early sixteenth century Ark, 7 (12), 405-15.

ARMITAGE, P.L. (1982) Studies on the remains of domestic livestock from Roman, Medieval, and early modern London: objectives and methods In (eds) A.R. Hall & H.K. Kenward, Environmental Archaeology in the Urban Context. C.B.A. Res. Rep. 43. London: Council for British Archaeology.

ARMITAGE P.L. & CLUTTON-BROCK J. (1976) A system for classification and description of the horn cores of cattle from archaeological sites J. Archaeol. Science 3, 329-48.

BARCLAY H.A. (1980) The Role of the Horse in Man's Culture London: J.A. Allen.

BINFORD L.R. & BERTRAM J.B. (1977) Bone frequencies and attritional processes In L.R. Binford, *For Theory Building in Archaeology* Academic Press.

BRAIN C.K. (1976) Some principles in the interpretation of bone accumulations In (eds) G. Ll. Isaac & E.R. McCown, *Human Origins* New York: W.A. Benjamin, 97-116.

CLARK J. (1980) Saint Erkenwald: Bishop and London Archaeologist *London Archaeologist*, 4, 1, 3-7.

CUTTING C.L. (1955) *Fish Saving* London: Leonard Hill.

DAVIS S. (1986) *The Archaeology of Animals* London: Batsford.

GRANT A. (1987) Some observations on butchery in England from the Iron Age to the Medieval period *Anthropozoologia* Premier Numéro Special, 53-58.

HOLMES J.M. (1981) Report on the animal bones from the Resonance Chambers of the Whitefriars Church, Coventry In C. Woodfield, 'Finds from the Free Grammar School at the Whitefriars, Coventry' *Post-medieval Archaeology* 15, 126-153.

JONES R.T & SLY J (N.D.) *The terrestrial vertebrate remains from the excavations at the castle, Barnard Castle* Ancient Monuments Laboratory Report No 4630.

JONES R.T. & SERJEANTSON D. (N.D.) *Animal bones from five sites at Ipswich* Ancient Monuments Laboratory Report 3951, 1983.

LANGDON J. (1984) *Horses, Oxen and Technocological Innovation* Cambridge University Press.

LYMAN R.L. (1987) On zooarchaeological measures of socioeconomic position and cost-efficient meat purchases *Historical Archaeology* 21, 1 58-66.

MACGREGOR A. (1985) *Bone, Antler, Ivory and Horn* London: Croom Helm.

MALTBY M. (1979) *The Animal Bones from Exeter, 1971-1975* Exeter Archaeological Reports 2. Sheffield: Department of Prehistory & Archaeology.

O'CONNOR T.P. (1984) *Selected Groups of Bones from Skeldergate and Walmgate* The Archaeology of York 15 (1) London: Council for British Archaeology.

PAYNE S. (1972) Partial recovery and sample bias: the results of some sieving experiments In (ed) E.S. Higgs, *Papers in Economic Prehistory* Cambridge University Press, 49-64.

RACKHAM O. (1986) *The History of the Countryside* London: Dent.

SCHIFFER M.B. (1987) *Formation Processes of the Archaeological Record* University of New Mexico Press.

SCHMID E. (1972) Atlas of Animal Bones Amsterdam: Elsevier.

SERJEANTSON D. (forthcoming) Fish remains from St Albans Abbey In Fishes and Mankind: Proceedings of the 4th I.C.A.Z. Fish Meeting, (ed) A.K.G. Jones.

SERJEANTSON D., WALDRON T. & McCRACKEN S. (1986) Veal and calfskin in eighteenth century Kingston London Archaeologist 5, 9, 227-232.

UERPMANN H. -P. (1973) Animal bone finds and economic archaeology: a critical study of osteo-archaeological method World Archaeology 4 (3), 307-322.

The Universal British Directory of Trade and Commerce (1791) London, 49-91.

VAN MENSCH P.J.A. (1974) A Roman soup kitchen at Zwammerdam? Berichten van de Rijkdienst voor het Oudheidkundig Bodermonderzoek, 24, 159-166.

WHAT SHALL WE HAVE FOR DINNER? FOOD REMAINS FROM URBAN SITES

Terry O'Connor

The purpose of this paper is to consider the means by which towns obtained their animal-based food supply during the past, and, to some extent, today. The implications of this production and trading system for the interpretation of animal remains from urban sites will be discussed, using examples drawn from work on material from York and Lincoln. Because archaeology, even in towns and cities, tends to be orientated towards individual sites, there is a tendency to forget the scale and complexity of past and present urban trading systems and to make a simple equation between the bones and shells recovered from a site and the diet of the population. The aim here is to show the extent to which the question 'What shall we have for dinner?' was constrained by factors not primarily concerned with the production of animal-based foods.

TOWNS AS MARKETS

Modern towns function on a sophisticated cash-based market economy, and it seems likely that a similar cash economy has typified towns in Britain since the Roman period. The development of trade specialisations within a town leads to the creation of a sector of the urban population who are not producers of food, and who thus require a mechanism by which to trade their services for a share of the food which others will have produced. Taxation must also have tended to act as a spur to the development of cash economies, for although a tax can be levied in the form of a tithe of grain or livestock, coins are more convenient, of more readily agreed value, and more easily stored in quantity. Although coins were certainly minted and circulated in pre-Roman Britain, it seems likely that cash marketing really began with the establishment of towns in the 'Romanised' districts of Britain. Rivet (1969, 200) proposed a model in which military pressure on local food markets created a cash trade in farm produce, and Hopkins (1980) has cogently argued that the imposition of taxes would have extended cash trading beyond the immediate sphere of influence of the Roman military. Thus from the 2nd century onwards, towns in Britain have largely been freed of the necessity of producing food provided that sufficient cash could be generated to buy it.

Some connection may be seen between the theoretical emergence of a market economy as outlined above and the bone debris from Roman urban sites. Maltby (1984) has suggested that deposits indicative of large-scale beef carcass processing represent trading in cattle on a scale which would only be consistent with a market economy, and gives examples of such deposits from military and civilian Roman contexts from a number of different towns. To these may be added recent results from late 2nd century deposits in York (O'Connor 1988), which show highly-specialised carcass processing being undertaken on a large scale. Typically, such deposits comprise substantial dumps of cattle femora, radii, humeri and tibiae, usually with relatively few epiphyses, and always with the diaphyses chopped and smashed into pieces a few centimetres in length (see also Maltby, this volume).

From the beginnings of urbanisation in Britain, then, towns have been markets for many commodities, including farm livestock and wild game. As towns expanded, so the catchment areas from which they obtained their food expanded. Medieval documents contain many examples of the movement of large numbers of farm livestock around the country. One of the best examples is quoted by Trow Smith (1957 110) which tells how in May 1323, one John Le Barber moved a total of 33 cattle and over 1500 assorted sheep and lambs between the Fens and Tadcaster, in North Yorkshire. Some parts of England appear to have traded cattle through receiving depots. One such centre at Macclesfield, Cheshire, has been well-documented for the years 1356-7 (Hewitt 1929, 50-54), and appears to have acted as a clearing-house for cattle from most of Cheshire, and also substantial parts of North Wales.

The implications of this extensive trading for our studies of bones from urban sites are considerable. A sample of cattle or sheep bones from a medieval context in a particular town will not represent livestock which have been raised locally as part of subsistence agriculture. For the most part, the animals will have been bought at local markets to which they may have been driven from a number of different, and probably distant, sources. Little wonder, then, that in biometrical studies of urban bone assemblages, within-sample variation often greatly exceeds that between samples.

ANIMALS AS FOOD

The supply of animal-based foods to towns in the past can be seen as deriving from three principal sources. The majority probably came from farms and estates, agricultural enterprises which may have been far removed from the urban market. A smaller proportion of meat, eggs and dairy produce was probably obtained from livestock kept in and around the town itself. Although it is customary to think of medieval towns as crowded places, early topographical accounts and the evidence of archaeological excavation make it clear that substantial areas of waste ground, orchard and 'garden' existed within even the large cities. The third source of meat was hunting or fishing, whether the opportunistic bagging of small game or the more organised fishing of coastal and inland waters. These three sources would have functioned independently of one another, as three discrete supply networks contributing different components to the food supply, and thus to the excavated bone sample.

Although this volume is primarily concerned with animal remains, any discussion of the supply of cattle and sheep to towns must take into account the growing of cereals and other field crops. Cows and oxen were the tractors of Roman and medieval agriculture, and however strong the market emphasis on cereals may have been in a particular region, plough teams had to be maintained, together with breeding-stock to replace worn-out draught cattle. This is not to suggest that beef and dairy produce were mere by-products of little worth. Meat, hides and milk products clearly had a substantial market value, though cattle appear to have been considered inferior to sheep and goats as dairy animals before about the 13th century (Whitlock 1965, 106). However, a consistently high proportion of cattle bones amongst those obtained from an urban site need not indicate deliberate stock-rearing for beef in the immediate locality. Intensive arable cultivation could produce the same effect, although it may be simplistic to suggest that Roman or medieval agriculture in Britain was sufficiently specialised to justify the use of terms such as 'arable farming' or 'cattle breeding'.

Sheep, too, had their place in the crop-raising economy, as a convenient means of cleaning up the harvest aftermath and spreading pelleted manure on the fields. Wool was probably the primary product of the sheep over most of the country, although in some areas a dairy industry based on sheep was flourishing by the end of the medieval period (Trow Smith 1957, 193). Sheep, like cattle, were not primarily kept for their value as meat until rapid population growth and a change in market demands revolutionised stockbreeding in the late 18th century.

THE MARKET FOR CATTLE AND SHEEP

Cattle and sheep bones comprise the great majority of most urban archaeological bone assemblages, and thus beef and mutton clearly constituted the bulk of the red meat consumed in Roman and medieval towns in England. However, if we consider the workings of the pre-mechanised farm as revealed in contemporary documents, it is clear that neither of these species was raised solely for food. The cattle and sheep which came to market in the towns, and whose bones ended up in rubbish pits and middens, were a product of mixed farming, and, increasingly through the 11th-14th centuries, a burgeoning wool industry.

Two archaeological examples will illustrate these points. In Figure 1, a simple comparison is made of the incidence of sheep and cattle bones in archaeological deposits from York and Lincoln. In both cities there is a progressive increase in the ratio of sheep:cattle bones; that is, sheep bones become more abundant relative to those of cattle. The rate and extent of the change differs a little between the two cities, but the same underlying process is being observed. Given the well-documented rise in the importance of wool as an export commodity (Lloyd 1977), this replacement of cattle by sheep in the diet can be seen as the urban food supply being modified by economic factors pertaining to another farm product, not simply by constraints of food supply and demand. In other words, the constituents of the bulk of York and Lincoln's red meat supply through the medieval period may have had less to do with what the citizens wanted to eat, and more to do with what carcass surplus the mixed farming economy of Eastern England was producing as wool and corn prices fluctuated.

Some support for the notion that cattle were not primarily raised for meat or for dairying may be seen in the fact that cattle in 9th to 15th century deposits from York and Lincoln show a distribution of age at death atypical of either production regime. Fitting calendar ages to observed dental eruption and attrition data is notoriously difficult, and any reconstruction of age distribution in cattle mandibles must be approached with caution. However, the great majority of specimens are usually from adult cattle, though not of any great age: the lower third molar is in wear, but not heavily worn. It is also usually the case that the later-fusing limb bone epiphyses are mostly fused, though the vertebral centra are often unfused. Converting this into approximate ages, through the medieval period in both cities it was apparently unusual to slaughter cattle before their fourth year, and unusual to keep them beyond their eighth or ninth year. A dairy herd would be expected to produce a surplus of male calves for slaughter as veal. As Table 1 shows, such a pattern of slaughter has been recorded in late - and post-medieval deposits in York, though not before. Trow Smith (1957 70) attempts to reconstruct the typical range of ages of a plough-ox team around the time of Domesday. His conclusions

	P	J	I	S	A	E
Coppergate late 9th-early 11th cen.	3	3	16	73	158	43
Coppergate 12th-13th cen.	0	3	2	5	33	10
Aldwark 15th-16th cen.	0	4	1	1	16	1
Aldwark 17th cen.	0	16	0	0	16	0

Table 1. Age grouping of cattle mandibles from 16-22 Coppergate and 1-5 Aldwark sites, York. The 17th century group from Aldwark includes numerous veal calves, probably indicating the beginnings of cattle-based dairying within the city's catchment area.

P = perinatal
J = juvenile: not newborn, but first permanent molar not in wear.
I = immature: first molar in wear, but second molar not in wear.
S = subadult: second molar in wear, but third molar not in wear.
A = adult: third molar in wear, but not heavily worn.
E = elderly: heavy wear on third molar, i.e. dentine exposed on the distal column.

would give a plough-ox population predominantly composed of four to eight or nine year-old beasts, so the age distribution of cattle seen in medieval samples would not be inconsistent with the majority having been obtained from mixed farms as stock surplus to, or retired from, the plough team. The increased use of cattle for dairying, and thus the appearance of veal in the urban meat supply, towards the end of the medieval period has been detected in bone samples from York (Table 1), where an increase in the proportion of young calves (stage J) can be seen in the Aldwark samples. It has also been noted in Exeter (Maltby 1979, 32) and London (Armitage 1983). Whitlock (1965 106-7) links this trend, somewhat ingeniously, with the increased use of horses for haulage, suggesting that the cow '...had to find a new economic justification for its existence'.

LIVESTOCK IN TOWN HOUSEHOLDS

Closer to home, and perhaps more controllable by urban demand, would have been the livestock kept on orchards, smallholdings and backyards immediately around and within the town. This source of food may have been independent of the network of production and trade based on farms in the rural hinterland, and its constituents may thus give a better impression of the tastes and needs of the urban population. Not all livestock can be kept in a town, of course. Apart from the odd dairy cow or ewe, it seems unlikely that Roman or medieval towns supported populations of cattle or sheep. However, goats, pigs, fowls and geese are all amenable to being kept in fairly restricted surroundings, fed on such diverse matter as urban refuse can provide and constituting a convenient supply of cheese, meat and eggs. For the urban householder, the goat must have represented a useful resource, with a long lactation period providing milk in modest but reliable quantities. There are difficulties with the interpretation of goat bones in urban contexts. Goat horn was a valued raw material in the past, and accumulations of goat horncores, often associated with those of cattle, are not uncommon in urban contexts (eg. see O'Connor 1984; MacGregor this volume). From York alone, five sites of 11th-13th century date have thus far produced concentrations of goat horncores. However, when considering goat remains in the context of food, it is desirable that the main carcass bones are quantified, which can only be done where separate identification of sheep and goat is made consistently on a wide range of skeletal elements.

The role of the pig as a device for converting garbage to meat, and thus as a convenient adjunct to the human household, is well documented (Dudley Stamp 1969, 151). A pig being fattened for the butcher can be kept in a relatively small space, although the breeding of pigs imposes greater requirements of space and shelter. (For a good account of 'low technology' pig rearing, see Blake 1956.) Although medieval practice in York and Lincoln would seem to have been to slaughter pigs only when they were at least a year old (O'Connor 1982, 33-4), York sites have produced small numbers of bones of foetal and neonatal pigs. This suggests that in 9th to 13th century York, at least, some breeding of pigs took place in backyards and open spaces around the city. The absence of young piglet bones from Lincoln sites is probably an artefact of recovery: the great majority of specimens from York are recovered by sieving soil samples, and this approach has not generally been adopted on excavations in Lincoln. The routine use of sieving on sites in York has also made clear the amount of fragmented egg shell, most of it apparently from hens' eggs, which was incorporated in urban deposits at all periods. The extent to which eggs were used in medieval cuisine, in particular, tends to be underestimated. Sources such

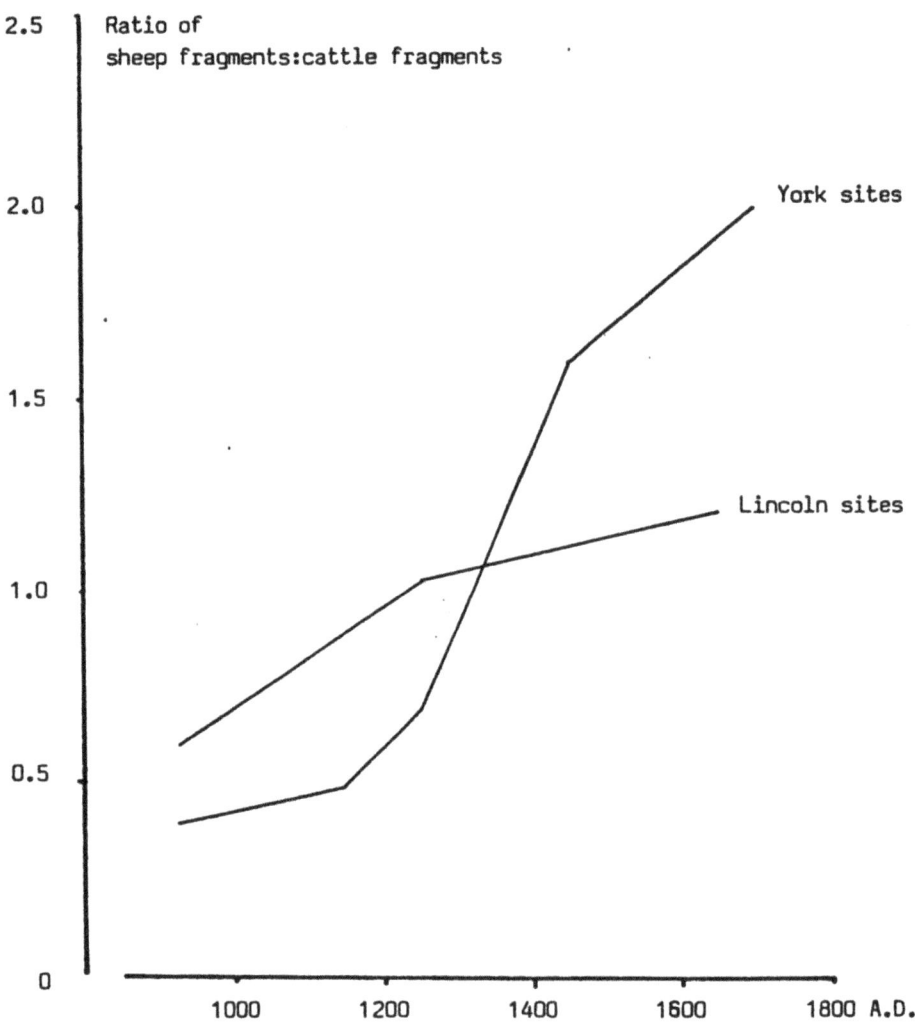

Figure 1. Graph to show the marked increase in sheep bones compared to cattle bones at sites in York and Lincoln dated between the 10th and 18th centuries. The data are derived from large assemblages of unspecialised urban debris. The procedure used, a simple comparison of counts of identified fragments, can be criticised on methodological grounds, but it is adequate to show the same relative increase in sheep bones in both cities.

as Labarge (1965, 81-2) detail a level of egg consumption which seems frankly staggering to modern eyes. Domestic fowl were thus important to the diet, though not primarily as meat, and the yards and orchards which housed urban pigs probably sustained poultry as well. 10th - 11th century deposits in York have also yielded numerous specimens of honey bee (Apis mellifera), a reminder that this animal was important as the producer of honey to sweeten food, as well as of wax. The city and its immediate environs thus provided a second source of food, probably producing a different suite of resources to the farms, and better placed to respond to local needs and demands.

HUNTING AND FISHING

The third source of animal based foods in Roman and medieval towns can be described as hunting, though hunting-gathering might be a more accurate term. It is in this category of animal-based foods that the urban cash economy will have been most effective. Whereas with a rural subsistence community the selection of wild species which can be exploited will be limited by local availability, the urban market could draw in resources from a much greater area, and thus a greater diversity of habitats. The limitation on which species of game, wildfowl and fish were offered for sale in Roman and medieval towns was not so much what was available as how much the citizens were prepared to pay for a particular product. To use a contemporary analogy, both potatoes and black olives are sold in York today. The olives are imported and thus expensive, but the demand is sufficient that the market will support the necessarily high price, and it is thus worthwhile for grocers and delicatessens to import them. Potatoes are locally available and cheap: if the price were to be raised to a level comparable with that of olives, few people would buy them. The controlling factor is the value which the urban market places on the product. If a high enough price could be commanded, the extensive trading links of Roman and medieval towns could provide the consumer with almost anything. To some extent, then, the diversity of hunted species consumed in a town at a given period could be taken as an index of wealth.

York is probably typical of English medieval towns in that hunted resources represent only a minor element in the food supply. Only certain species of fish, notably herring (Clupea harengus) and eel (Anguilla anguilla), seem to have been consumed in sufficient quantities to be seen as part of the staple diet. Wild mammals and birds are consistently present in archaeological deposits in the city, but in such small quantities as to indicate that they were only used as an occasional variation in the diet, and certainly not as a resource which was relied upon. Changes in the range of fish species consumed in the city between the 9th and 13th centuries show a progressive widening of the catchment from which fish were obtained (A.K.G. Jones pers. comm.). In the 9th and early 10th centuries, the exploitation was fairly local to the city, concentrating on river and estuarine species. Small cyprinid river fish were eaten in quantity, despite their somewhat questionable palatability, and eels were an abundant, probably locally available resource. From the mid-10th century onwards, herrings figure ever more commonly in the food debris, to be joined in the 11th and 12th centuries by cod (Gadus morhua) as the fishery extended into inshore waters. Later in the medieval period there is evidence of a further extension, as haddock (Melanogrammus aeglefinus) and ling (Molva sp.) bones become more common. Over the course of five centuries, then, the exploitation of fish in York changed from opportunistic catching of whatever

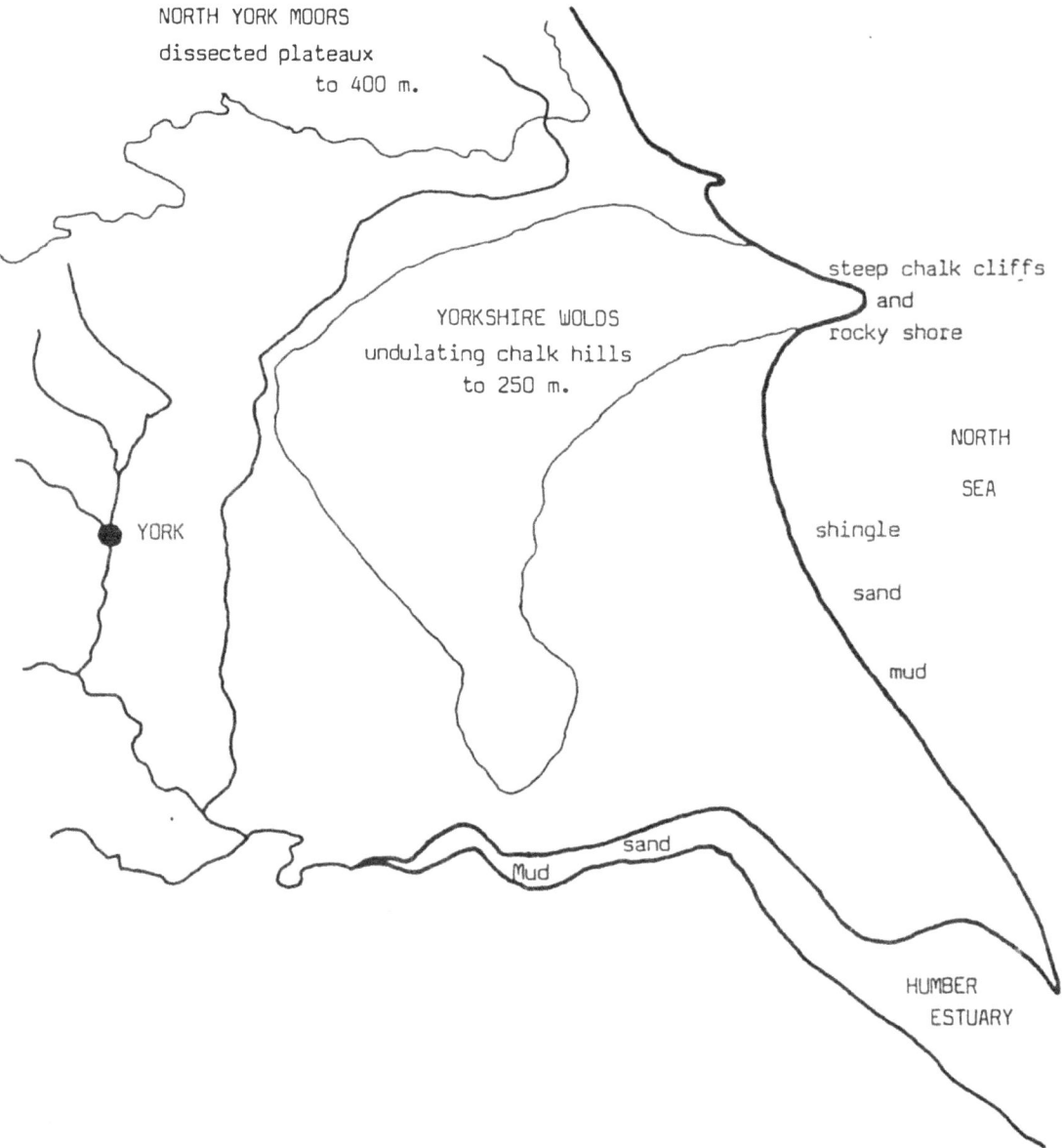

Figure 2. Sketch map of East Yorkshire to show the relationship of York to coastline and major rivers. The steep, rocky shores formed where chalk outcrops at the coast provide suitable habitats for edible crabs, imported to York in the Roman period, and auks, imported during the 11th - 13th centuries. As fishing boats from York probably exploited the Humber estuary for oysters, mussels and inshore fish, it is uncertain whether the crabs and auks were transported overland or by sea and river.

could be had in the local rivers to an organised trade in fish from North Sea waters.

By contrast with the fish, the exploitation of wild birds and mammals in York shows little major change between the second and 15th centuries, with duck and goose species, hare (Lepus europaeus) and roe deer (Capreolus capreolus) being eaten at all periods. One minor change in the exploitation of coastal resources may, however, serve as an example of market demand rather than availability affecting the food supply of the town. Amongst the diverse coastal and estuarine habitats within the catchment area of York, the East coast between Flamborough Head and Filey has extensive areas of exposed rocky shore backed by steep chalk cliffs (Figure 2). The rocky shore supports populations of edible crab (Cancer pagurus), whilst the cliffs provide nesting habitats for a variety of seabirds. Crab shell is a common find in soil samples of Roman date: in a recent study of 2nd-3rd century deposits, crab shell was identified in about one-third of all soil samples (O'Connor, 1986a). Clearly there was sufficient demand for this commodity in the Roman town to justify transporting crabs from the coast, either by an overland journey of around 50km or by travelling well over twice that distance by sea. In medieval deposits, however, crab shell is scarce, and it is clear that there was much less demand for the product, insufficient to support a regular trade, despite substantial exploitation of inshore fish and oysters (Ostrea edulis). Instead, 11th - 13th century deposits contain the bones of another food resource from the same stretch of coast, namely guillemots (Uria aalge) and razorbills (Alca torda). Within the last few centuries, there has been a tradition of exploiting seabirds for meat, eggs and feathers in East Yorkshire, and it seems likely that this resource was always important to coastal communities (Allison 1985, 181-2). It is only during a relatively short period of perhaps 200 years, however, that auk bones are found in archaeological deposits in York. Evidently there was a demand for this particular delicacy for a while, but it is notable that the trade in seabirds was not accompanied by a trade in crabs from the same stretch of coast.

CONCLUSIONS

Animal-based food remains from urban sites must be seen as the product of a complex of interlinked but discrete economic systems. Different products were obtained from different sources, with market demand determining to some extent what came into the towns, and constraints linked to the requirements of wool and arable production also having an effect. As the bulk of the meat consumed in Roman and medieval towns in lowland England appears to have been obtained from cattle and sheep, it can be argued that the population of a particular town could only directly influence the minority component, that obtained from stock kept immediately within and around the towns and from hunting and fishing. From the 12th century onwards, and perhaps also during the 3rd-4th centuries (Hoskins 1955), the number of sheep and cattle available in the rural economy to feed the towns was determined to a large extent by the woollen industry, arguably the engine which powered the economy of medieval England. Changes in the relative proportions of these two species observed in urban bone debris are thus likely to provide information on a regional scale rather than a local one. Spatial heterogeneity in the distribution of distinctive 'suites' of cattle and sheep bones within a single phase of urban development, on the other hand, may allow discrimination of high and low status areas, industrial quarters from residential areas, and so on.

It is in the changes within the 'smallholding' and hunted species that most information directly relevant to the town will lie. Social stratification may be apparent if certain parts of a town show a markedly greater reliance on locally available, perhaps 'home-grown', produce than others. The purchasing power of districts of a town, or even of individual households, will be shown clearly in the range of hunted and fished species exploited (see Ijzereef, this volume). Although the species represented will to some extent reflect the surrounding environment, the ability of urban markets to draw in goods from far afield must be borne in mind. The recent discovery of garden dormouse (Eliomys quercinus) bones in Roman deposits in York has served as a timely reminder of how far exotic species may be deliberately or accidentally transported (O'Connor 1986b).

The remains of animals exploited for food are often abundant on urban archaeological sites, and their potential as sources of socio-economic evidence is considerable. Long-distance trade and interacting market forces will produce many pitfalls, however, and a straightforward interpretation of results will be either unwise or impossible. Perhaps the most pressing need in urban bone studies is to be able to model the urban economy in sufficient detail to be able to predict what bone debris might have fallen out of it at different times and places.

REFERENCES

ALLISON, E.P. (1985) _An archaeozoological study of bird bones from seven sites in York._ D. Phil. thesis, University of York.

ARMITAGE, P.L. (1983) Faunal Remains, In Thompson, A., Grew, F. and Schofield, J, Excavations at Aldgate 1974. _Transactions of the London and Middlesex Archaeological Society_, 129-150.

BLAKE, N. (1956) _The Pioneering Pig_, London: Faber and Faber.

DUDLEY STAMP, L. (1969) _Man and the Land._ London: Collins, New Naturalist Series.

HEWITT, H.J. (1929) _Medieval Cheshire._ Chetham Society Publication 88.

HOPKINS, K. (1980) Taxes and trade in the Roman Empire (200 BC - 400 AD). _Journal of Roman Studies_, 70, 101-125.

HOSKINS, W.G. (1955) _Sheep Farming in Saxon and Medieval England._ London: Wool Education Society.

LABARGE, M.W. (1965) _A Baronial Household of the 13th Century._ London: Eyre and Spottiswoode.

MALTBY, J.M. (1979) _The Animal Bones from Exeter 1971-1975._ Sheffield: University of Sheffield, Exeter Archaeological Reports vol. 2.

MALTBY, J.M. (1984) Animal bones and the Romano-British economy. In *Animals and Archaeology: 4 Husbandry in Europe*, ed. C. Grigson and J. Clutton-Brock. Oxford: British Archaeological Reports S227, 125-138.

O'CONNOR, T.P. (1982) *Animal bones from Flaxengate, Lincoln c870-1500.* Archaeology of Lincoln 18(1). London: Council for British Archaeology.

O'CONNOR, T.P. (1984) *Selected Groups of Animal Bones from Skeldergate and Walmgate.* Archaeology of York 15(1). London: Council for British Archaeology.

O'CONNOR, T.P. (1986A) *Mollusca and other shell fragments from the General Accident Extension site, York (1983-4, 32).* Ancient Monuments Laboratory Report no. 4768.

O'CONNOR, T.P. (1986B) The garden dormouse *Eliomys quercinus* from Roman York, *Journal of Zoology, London*, 210, 620-622.

O'CONNOR, T.P. (1988) *Bones from the General Accident Extension Site*, Archaeology of York 15(2). London: Council for British Archaeology.

RIVET, A.L.F. (1969) Social and economic aspects. In *The Roman Villa in Britain*, ed. A.L.F. Rivet, London: Routledge and Kegan Paul, 173-216.

TROW SMITH, R. (1957) *A History of British Livestock Husbandry to 1700*, London: Routledge and Kegan Paul.

WHITLOCK, R. (1965) *A Short History of Farming in Britain*, London: John Baker.

THE PROVISION OF FOWLS AND FISH FOR TOWNS

Jennie Coy

INTRODUCTION

This paper discusses the retrieval and analysis of bird and fish bones from recent archaeological salvage work in towns in Central Southern England. The Faunal Remains Unit (F.R.U.) was set up at Southampton in January 1975 as the Faunal Remains Project. During this time bird and fish bones have been studied alongside those of the domestic and wild mammals. Much of the material for study has come from towns and has included major collections from Roman Winchester, Saxon Southampton (Hamwic), the medieval towns of both, and medieval and post-medieval collections from Alton, Christchurch, Newbury, Poole, Reading, Romsey and Salisbury. Work is now beginning on material from Roman and medieval Dorchester.

THE RETRIEVAL OF URBAN FAUNAL SAMPLES

Sampling Strategies

All the bones discussed here have come from rescue excavations which, through urgency and uncertain funding, were often subject only to limited research design. Adequate sampling of all types of context available and the sieving of soil samples from every layer have proved essential for a proper understanding of the role of fish and birds. Both trowelling and sieving are needed to gain a full picture. Whereas the latter picks up small elements of large species and new small species, the former samples an enormous amount of soil by comparison. It is essential when designing a bone retrieval strategy to ensure that the volume of soil sieved can be expressed as a proportion of the total volume of the context. Results from these two different methods of excavating should also be separable in the archive.

Some contexts will require far less sieving than others. Finlayson and Bellhouse (forthcoming) have shown that for a representative sample of faunal remains and potsherds on a North American Iroquoian site 30%-35% of large middens need to be sampled, 40%-45% of medium-sized middens, and small middens should be totally excavated.

For urban deposits a number of factors will determine the amount of soil which needs to be sieved. These include the type of context and its taphonomic history (basically how factors have acted on it since deposition), the variability within a context, and the amount that is already known about the town from previous excavation. In a well-studied town where rescue is urgent and poorly funded an excavation may set out to answer only a limited range of questions and these questions should include questions of environmental importance if they are to be broadly-based. But it should be remembered that to set these questions in context there needs to be a comparability with previous work in that town or elsewhere and that if the sampling aims are too narrow, one can lose the ability to answer a range of environmental questions.

Environmental Strategy and Project Management

1. Before and During Excavation.

In view of all these factors it is obvious that the strategy may need to be adjusted as the excavations proceed, although it helps the excavator and assistants if there is a clear default strategy. One for birds and fish in Wessex has been to retrieve 5 litre soil samples from every layer of any feature excavated, or from an agreed sub-sample of features, with much smaller samples where material is waterlogged.

With consistent methods the work in a town becomes cumulative and the questions become quite different. In the early stages they might simply be: "What species are present?", "What size were the fish?", "Were the wild birds eaten?". More refined questions, such as "Was this household of high status?", will emerge when a variety of context types has been sampled. It is important to stress that, until the groundwork has been done in a town, refined questions cannot be tackled.

In Southampton in the past most of the intensive study of soil samples has been in the Middle Saxon town (Hamwic). These have covered a wide range of contextual variability and types of preservation so that refined questions are now possible (Bourdillon n.d. 1, & forthcoming). In medieval towns, however, variability is more complex as activities are more variable and more localised.

This is only one reason for bringing together the environmental specialists before excavation, however briefly, to discuss the strategy in advance. It is advantageous if trial excavation or assessments of some kind can take place in order to forecast more accurately requirements for the main excavation. During these stages it can be decided who will share samples. Many specialists prefer to take their own.

There are a number of pitfalls in sample taking which only the specialists or their contact on site may know and, despite the need to employ a standardised methodology over a period of time, we are all learning all the time. Any excavation team will benefit from even a brief contact with specialists as it will make their work more interesting and the excavation more relevant. This is a small extra cost which must be built into the overall cost of urban excavations if we are to gain new knowledge from them.

In large units it is necessary to have a full-time environmental liaison worker who ties specialist needs to day to day work on site. Environmental specialists, although varied, are only one group in today's complex urban archaeology and the bulk samples taken with fish and seeds primarily in mind often yield important information for the ceramicist, bone finds specialist, and experts on metal-working and other technologies. This was made strikingly clear during work on the Pit Project undertaken by Sarah Colley and IBM (UK) Research at Hamwic where the contents of a Middle Saxon pit were totally sieved through 1mm mesh and the position of thousands of artefacts recorded three dimensionally (Colley, Todd & Campling 1988; Colley, forthcoming).

With the specialist liaison and sieving and sorting arrangements set up, early evaluation of a few samples while the excavation is still on will lead to a more relevant sampling programme. Ideally, sieving and assessment of soil samples should keep up with the excavation. Although long term rescue units may have a fairly clear idea of what is needed and have a standard set of procedures for bulk sampling and processing (e.g. Cottrell forthcoming), on new sites or with new types of context it is impossible to lay down a detailed strategy until you can see what is coming up.

It may be decided that too much soil has been saved and that is should not all be sieved. This sort of decision is only possible when sieving is keeping up with excavation so that likely finds can be quickly scanned. It is so much easier to store processed flots and residues than bags of soil that an efficient sieving programme is essential. With some kinds of preservation, however, - waterlogging or mineralisation - decisions must be made early on how much to save as their processing and analysis can be far more labour intensive.

With a rapid throughput at the sieves more flots and residues may sometimes be produced that need to be studied in detail, and it is better to ask the specialist to do a quick scan than to ask untrained workers to spend hours sorting out what may be irrelevant.

The worst kind of urban archaeology for an understanding of the role of birds and fish is that which relegates specialists to the post-excavation phase.

2. Post-Excavation

The post-excavation stage itself needs careful planning and close co-operation between environmental specialists and others involved.

Bird and fish bones from different types of contexts may have quite different origins and it is important to know as much as possible about the likely formation processes involved in a particular build up of deposit before study of the bones. Once examined the bones may or may not fit this picture and can sometimes lead to reconsideration of the provisional archaeological conclusions for a context. The bones may provide evidence for the origin of the deposit not shown by the other finds, for example, that it contains cesspit material. They may show up evidence for residuality or contamination that either supports or does not support the picture from ceramics. Sometimes this work may need detailed discussion between the specialists about the stratigraphy and phasing.

A recent example shows how techniques must be adjusted to the site and its prime archaeological questions. In a study of soil samples from the medieval fish market at St. Michaels, Southampton, our deliberations were handicapped by our ignorance of what ordinary occupational debris in medieval Southampton should look like! (Coy & Hamilton-Dyer, forthcoming). Only a small amount of material from somewhat specialised contexts had ever received similar treatment in Southampton (Hillman, forthcoming, Coy, forthcoming) and the main bulk of samples from excavations on medieval sites still remains to be studied. As we were working in the dark we therefore looked at some very large samples, from two extensive contexts which were thought by the excavator to be contemporary levels of the fish market itself. Because they were extensive contexts several standard samples had been taken which aimed to sample any horizontal variation in the material.

Species	No.Fragments	Weight in Kg	MNI overall*	MNI Cumulative**
Horse	49	4.4	5	21
Cattle	23,896	587.9	211	422
Sheep/Goat(Goat)	14,606(130)	128.1(7)	265	480(59)
Pig	6,953	94.8	192	386
Dog	23	0.2	4	9
Cat	144	0.1	13	25
Fowl	800	1.4	63	199
Goose	353	1.3	16	102

* Minimum Number Individuals for Melbourne Street as a whole
**Minimum Number as a cumulative total of separate features

Table 1. Domestic animals from Melbourne Street, Hamwic.

	normal recovery	sieving
Dasyatis pastinaca, stingray	1	
Raja sp, mostly thornback		112
Anguilla anguilla, common eel		337
Conger conger, conger eel	5	42
Clupeidae, herring family (mostly herring)		349
Possible Cyprinid, carp family		1
Gadus morhua, cod	9	11
Melanogrammus aeglefinus, haddock	14	12
Merlangius merlangius, whiting		65
Trisopterus luscus, bib		4
Pollachius virens, saithe	2	
Pollachius species, saithe or pollack	1	
Molva molva, ling	7	1
Gadidae, cod family	11	35
Merluccius merluccius, hake		4
Belone belone, garfish		5
Atherina presbyter, sand smelt		21
Triglidae, gurnards		40
Dicentrarchus labrax, bass		1
Labridae, wrasses		1
Scomber scombrus, mackerel		8
Pleuronectidae, (plaice, flounder or dab)		128
Unidentified fish fragments	32	2828

Table 2. Fish fragments from St. Michael's, Southampton. Seven species were recovered in the trench, mostly haddock and the other large gadoids. At least 12 additional species were only recovered by sieving, including eel and herring, from which the greater number of bones were present.

After looking at a sub-sample we decided on the basis of a scan that to look at more would reveal no further information relevant to the immediate questions. But the balance of the sieved material remains if further resources allow a detailed examination of, for example, fish sizes, or if new light is thrown on these deposits by future examination of material from, for example, contemporary domestic contexts.

Our conclusions about the fish market are still tentative but this detailed work can now serve as a basis for any future fish work in medieval Southampton and has already led to an ability to 'scan' sieved material as we have done for some years for the common ungulates (Bourdillon n.d.2, n.d.3, compare O'Connor, this volume). Such sub-sampling and scanning can cut down the amount of labour-intensive sorting, especially where it involves picking out small fish bones under a binocular microscope, but it must be organised by structured sampling techniques and involve the specialist closely.

THE EVIDENCE FOR FOWL IN TOWNS

The Collins dictionary defines 'fowl' as 'domestic fowl' or 'any other bird (esp. any gallinaceous bird) that is used as food or hunted as game'. I shall concentrate mainly but not exclusively on the domestic fowl itself and the importance of its remains in archaeological interpretation generally. In connection with 'game' I shall make some reference to falconry.

The Rise in Importance of Domestic Fowl

The common domestic fowl or chicken (Gallus gallus), thought to be a domestic form of one or more of the Asian jungle fowl, is the most common bird find in urban sites in Wessex. There is a little evidence of fowl in Wessex in the Late Iron Age, for example, from the Late (b) phase at Danebury (300-100/50 B.C.), and it is presumed to have been introduced from mainland Europe (Coy 1983i; Coy 1984; Harcourt 1979; Maltby 1981).

By the 1st and 2nd centuries AD in Wessex it provides over half the bird bones found. Domestic fowl averages out at 60-75% of bird bones from Roman, Saxon, and early medieval Winchester deposits.

But the number of domestic fowl fragments is not usually as high as the number of domestic ungulate fragments. Table 1 shows this for Melbourne Street, Saxon Southampton (Bourdillon & Coy 1980). It also shows the ratios of the main species among the material preserved. At this site Bourdillon calculated that at least 96% by number, and perhaps as much as 99% by weight did not survive (Bourdillon 1979).

Domestic Fowl Bones as a Methodological Tool

The remains of fowl on sites may be used to set up a model for investigating disposal strategies as an alternative to the more commonly used ones involving ungulate bones (e.g. Maltby 1985). This has some advantages which are detailed below:

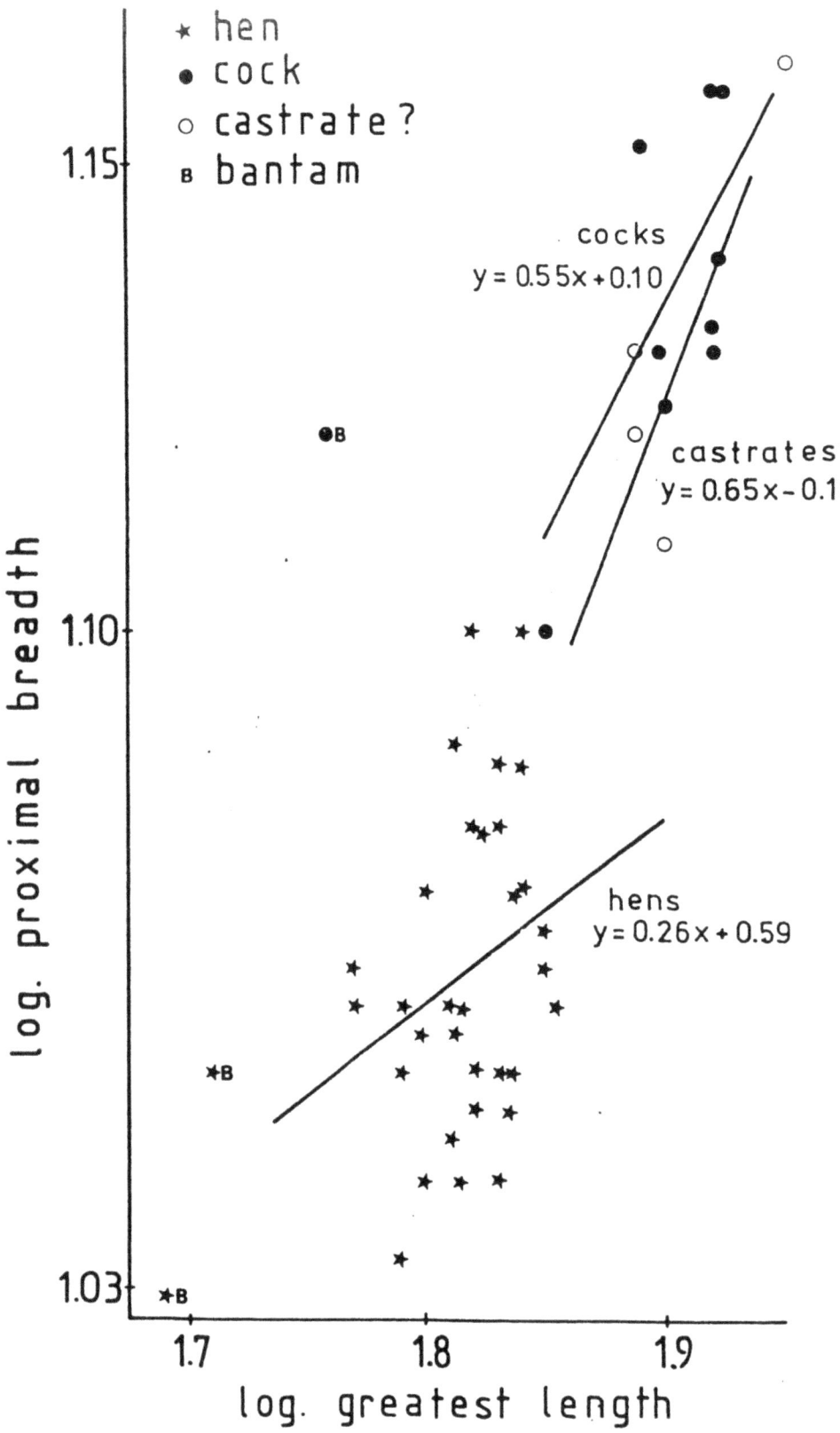

Figure 1. Size groups of domestic fowl from Hamwic. Log greatest length of the tarsometatarsus is plotted against log proximal breadth (from Coy, 1983i).

1. Identifiability to Species

Provided that the presence of pheasant (e.g. Phasianus colchicus), guinea fowl (Numida meleagris), turkey (Meleagris gallopavo), black game (Lyrurus tetrix), capercaillie (Tetrao urogallus), and peacock (Pavo cristatus) can be excluded (and for this it is necessary to have access to a good modern comparative collection) a higher proportion of fowl bones can be identified to species than ungulate bones.

This phenomenon is caused by the relatively small size of the fowl bones and their lack of chopping, butchery normally being in the form of knife cuts. (They are however more fragile than bones of ungulates and when the fragments of bird bone from sieving are included quite different numbers will be obtained.)

2. Measurements

Provided then that the bird bones are separated as soon as possible from other finds they are more often whole so that comparative sizes and statures are more readily obtainable for a comparison of sites and periods. Sizes *do* change, however; for example some Roman and Post-medieval layers provided very large chicken bones.

The study of changes in fowl size is in its infancy compared with the large amount of work on the changes through time of the common domestic unuglates. Log/log graphs, which have been used in Figure 1, are probably most useful for monitoring size variations at present but as we acquire large databases more sophisticated statistical methods may be useful. Change in size in domestic animals are complex, with increases in size being due to an amalgam of heterosis (hybrid vigour) and changes in care and feeding, as well as selection. For example, some fowls were comparable with modern bantams at Hamwic (Figure 1) (Coy 1983i; Bourdillon & Coy 1980).

Work in Wessex seems to suggest that there are periods of size change in fowls which may mirror ungulate size changes. One way ahead may be closer analysis of ranges and maxima as an indicator of heterosis. Changes in proportion might also be worth monitoring e.g., size of legs could be a guide to function - layers short, fighters and sitters long. The sternum might be the best bone to monitor as an indicator of selection for meat value but it is often not preserved.

3. Carcase Treatment and Distribution

Individual fowl are more likely to have been reared, slaughtered, utilised, and discarded within an individual property than individual ungulates. This is more significant for taphonomic analysis for medieval and post-medieval deposits, when many of the other species may not give this kind of evidence. The small size of the fowl carcase, compared with that of an ungulate, reduces the amount of distribution or exchange that is likely to have occurred. Division of the carcase will be less likely, as will butchery through individual bones. Taphonomically fowl tell us about different post-depositional events from cattle bones and will therefore give different information on the formation processes of contexts.

Some of the individual bone elements show different possibilities for survival and distribution. The principle bones of domestic fowl are shown in Figure 2. High values for humerus and ulna may often be a sign of a high degree of residuality (Coy 1983i). The wings have little meat on and may have been removed from the carcase before presentation, especially if they were damaged, as in some trussed birds in supermarkets today. In this case the carpometacarpus would also be discarded. It is readily recognizable; however, because of its smaller size it is less likely to be recovered in trowelling than the humerus and ulna.

The coracoid on the other hand is a sturdy and distinctive bone and is more likely to provide an index of what happened to the rest of the carcase. The femur is a good meat bone and important for analysis of egg-laying as described below. The attached tibiotarsus or 'drumstick' provides a convenient and manageable snack illustrated by the current fashion for turkey drumsticks. Not only is it the largest long bone of the body but the distal end of the bone provides a handle for eating and pitching. All these factors could explain differences in deposition which as yet remain unstudied. Urban Hanwic showed significantly higher values for leg bones than some other Saxon settlements studied in the same way (Coy 1983i).

The tarsometatarsus, important for sexing the birds, may sometimes have been removed before cooking, as it is on some species today. It could in some periods therefore give evidence of food preparation. Sieving is very important for investigating food preparation. This is brought out by common finds from some deposits of cervical (neck) vertebrae and tracheal (windpipe) rings retrieved on the sieves, and the digits, both from wings and feet, are mainly recovered in sieving.

4. Evidence for Breeding

There is documentary evidence for backyard fowl breeding from medieval towns (Keene, 1985). A very large number of fowls can be reared in a backyard if there is proper accommodation for them and plenty of nearby plant material for them to forage in or for gathering. Detailed knowledge of fowl-rearing is documented from Roman times (e.g. Columella, Res rustica VIII) and medieval texts appear to be firmly based on Roman ones. Free-ranging fowl were certainly regarded as a hazard by medieval Winchester corn merchants (Keene 1985).

The major archaeological evidence for backyard urban fowl rearing is the occurrence of bones of immature fowl and of whole, unbutchered skeletons from casualties. Breeding status can sometimes be distinguished amongst the latter from the presence of medullary bone in laying hens.

The possibilities of sex distinction from sexual dimorphism, tarsometatarsus, and medullary bone have been discussed elsewhere, as has evidence for caponisation (Bourdillon & Coy, 1980; Coy, 1983i). Caponised fowl were used for the pot as well as for raising chicks.

Eggs survive on some sites. In recent work on the Southampton medieval fishmarket small fragments of eggshell were retrieved on the sieves from many contexts. It is possible sometimes to distinguish the species to which fragments of eggshell belong using electron microscopy (Keepax, 1981).

From the Medieval Period both fowl and eggs are likely also to have

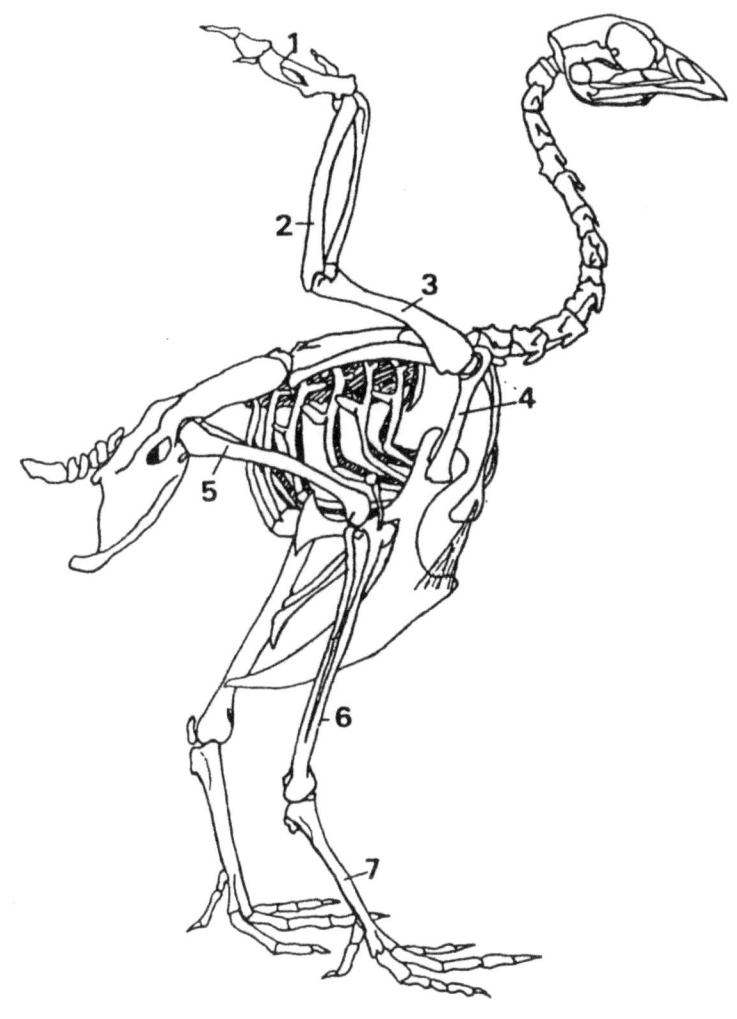

1 Carpometacarpus
2 Ulna
3 Humerus
4 Coracoid
5 Femur
6 Tibiotarsus
7 Tarsometatarsus

Figure 2. Skeleton of the domestic fowl, showing the bones referred to in the text and other principal bones.

come from local distributors and shops. The medieval Winchester poultry trade certainly traded in geese, hens, capons, partridges, and eggs (Keene, 1985). In addition the cooks bought live fowl and were supposed only to sell them cooked, probably often in pies. Records for Winchester show that they frequently broke the law and sold fresh meat and that they also sold pigeons, partridges, pheasants, plovers, woodcocks, and geese (Keene, 1985).

5. Additional Variability

Fowl bones provide an additional source of variability in multidisciplinary studies aimed at finding the differences, e.g., in standard of living, between different parts or periods of a town. In the past their presence in food remains would itself indicate high status and aditional information may be available on consideration of their age, numbers, and sex. But in order to work out whether there were consumer and producer sites for fowl, it is essential to distinguish between the remains of dead chickens and the remains of food by looking carefully for evidence of butchery. In large samples this might involve a spot check on each proximal humerus and/or each distal tibiotarsus for knife cuts as it is measured.

Styles of butchery may change in late medieval or post-medieval times when the tibiotarsus may be chopped through or birds halved, but in Wessex this tends to be rare although no systematic analysis of our records of fowl butchery has taken place since the close examination of the first large Hamwic sample. There, scrapes with a blade suggested knifemarks made removing meat to mouth (Bourdillon & Coy, 1980).

Other Gallinaceous Birds

In Wessex there have been a number of finds of introduced species related to the domestic fowl. Peacock bones are an occasional find in medieval towns, e.g. a peacock carcase from medieval Southampton was determined by Bramwell (Platt & Coleman-Smith, 1975). Pheasant has not yet been determined from a medieval urban site in Wessex although it has turned up in post-medieval rural contexts. This is surprising as it is occasionally mentioned in medieval documents. The only turkey bones found in an urban context in Wessex so far came from contexts dating to the Seventeenth or Eighteenth Century A.D. in Christchurch (Coy, 1983ii). The most reliable and often quoted reference to turkey in the British diet is in Cramner's 'A Dietarie' in 1541 (Gurney, 1921). Guinea fowl has not been recorded so far in Wessex, but it can be difficult to distinguish some of the bones from those of domestic fowl.

The native blackgame, which survived in Hampshire into the twentieth Century (Nicholson 1929) has occasionally been found and there are many medieval references to the much smaller partridge, Perdix perdix, whose bones are not uncommon in Roman, Saxon and Medieval urban deposits.

Goose Remains from Towns

There is a problem of distinction from bones of the probable wild ancestor, the greylag goose, Anser anser (Bourdillon & Coy, 1980). Occasional such large goose bones, probably from domesticated forms, are found from early phases of Iron Age settlements in Wessex (e.g. Harcourt, 1979; Maltby, 1981; Coy, 1984). It seems that domestic geese predated domestic fowl. Large greylags or domestic geese are found in Roman deposits in Wessex and Saxon geese are ubiquitous in urban middle Saxon Hamwic and must have been about the town, perhaps with their minders as they commonly are in Polish villages today. The wild species breeds in the North and northern Europeans were probably to the fore in goose cultivation. Riddell (1943) stresses the ease with which they might be caught while moulting their flight feathers.

He also gives an interesting account of the many uses of the domestic goose. Besides the importance of their eggs, down, and occasional meat, the best quill pens were made from primary quills plucked from live geese in the Spring and goose wing feathers would be important for arrow flights. A horrifying picture of the life of Eighteenth Century geese is shown by their annual value of "1s to 16d annually by being four times plucked" (Ernie, 1961). A city ordinance of 1380 banned geese from the main streets of Winchester (Keene 1985).

Duck Bones from Towns

There is a problem in distinguishing domestic duck bones from those of the mallard, Anas platyrhynchos, which is widespread in Britain. Anatomical differences which may be associated with domestication have now been seen from several Late Saxon collections including, for instance, Northampton (Coy, 1981).

Ducks are not mentioned in any of the medieval treatises on farming though they appear in the Berkeley accounts in 1321 (Smyth, 1883). But these accounts are for manors and it is not clear what happened in medieval towns. The city ordinance mentioned above for Winchester also applied to ducks.

Other Birds

Dorchester excavations produced the lower courses of a medieval dovecot for which there is a Fourteenth Century documentary reference (Susan Davies, personal communication). More usually associated with manors rather than towns, pigeons could provide an important food source and the young birds were eaten.

There is also evidence that people in medieval towns kept hawks and falcons. These provide important evidence of the hunting of birds for food and some of the wild birds found in town sites are those species commonly caught by, for example, the goshawk, Accipiter gentilis. Falcon bones are difficult to distinguish but an exciting recent find from Twelfth or Thirteenth Century Winchester was the partial skeleton of a large falcon which only really matches the gyrfalcon, Falco rusticolus (Coy, n.d.1). This is the largest and most prestigious medieval falcon and was probably imported for the nobility (if not royalty) from Scandinavia. There is some interesting documentation that might link this area of Winchester with the Royal Mews (Keene, 1985).

A wide, but not surprising variety of other species of bird have now been determined in Wessex towns and a summary of Saxon and Medieval results for Southampton will shortly be published (Coy, forthcoming). In total, Hamwic produced over twenty species of bird and medieval Southampton thirty-nine.

FISH REMAINS FROM TOWNS

Fish results for urban sites, even more than bird results, are heavily reliant on good retrieval; if sieving is not carried out only the bones of large fish are recovered and very biassed results are obtained. Table 2 shows the species list for Southampton medieval fishmarket from normal recovery against that for sieving (Coy and Hamilton-Dyer forthcoming). The smallest specimens may have been from food fish or may be the gut contents of larger fish but in that case are useful indicators of possible fish gutting activity. In 1427 a group of fishmongers in Winchester were accused of depositing the rotting entrails of fish on the Highway at Staple Garden at the rear of their premises (Keene, 1985).

Sieving can produce large quantities of fish bones and a wide variety of species and a good modern reference collection is needed to identify them. Modern specimens, if weighed and measured, can then provide additional data for the ancient bones if measurements are taken on these too. Regression methods enable us to assess the size of the ancient fish from as little as one standard measurement on a single bone. Figure 3 shows a graph obtained for modern eels where total length is set against the chord length of the cleithrum bone. The total length of archaeological eels can then be assessed by measuring cleithra, which are better preserved than most head bones.

Table 2 shows the relative importance of eel and herring among the fish, and this was the case in Wessex at least from Middle Saxon times onwards. The pickling, drying, and salting of fish has been important since prehistory and so one should record which parts of the skeleton are present as this may provide evidence for preservation processes. Marine fish, often of considerable size, are commonly found in medieval towns a long way from the sea. By the Medieval Period considerable trade had developed as we know from documents. Fishmongers from far afield, including London, traded in Winchester by the Fifteenth Century (Keene, 1985).

Records for the port of Southampton from 1300 show the wide range of shipping which brought both fresh and preserved fish and also show what went out of the town to nearby ones. Fourteenth Century records refer to prices levied on the movement of red and white herring, sardine, conger, cod, ling, fresh and salt mackerel, stockfish (cod and related species preserved by drying and salting), barrels of mulwell (preserved cod) and haddock, baskets of lampreys, barrels of sturgeon, and salt and fresh salmon (Studer, 1910). Fifteenth Century records include hake, pollack and ling from the South West, herring and sprats from Suffolk and Norfolk and whiting from Dieppe (Studer 1913). A daily record survives for many of the years in the Fifteenth and Sixteenth Centuries for some tolls and gives a picture, for example, of overland travels for herring, salmon, and hake (Coleman, 1960).

The fascinating but incomplete picture give by such documents and those which record court cases make us more wary of seeking local explanations for changes in fish exploitation through time, whether ecological or technological.

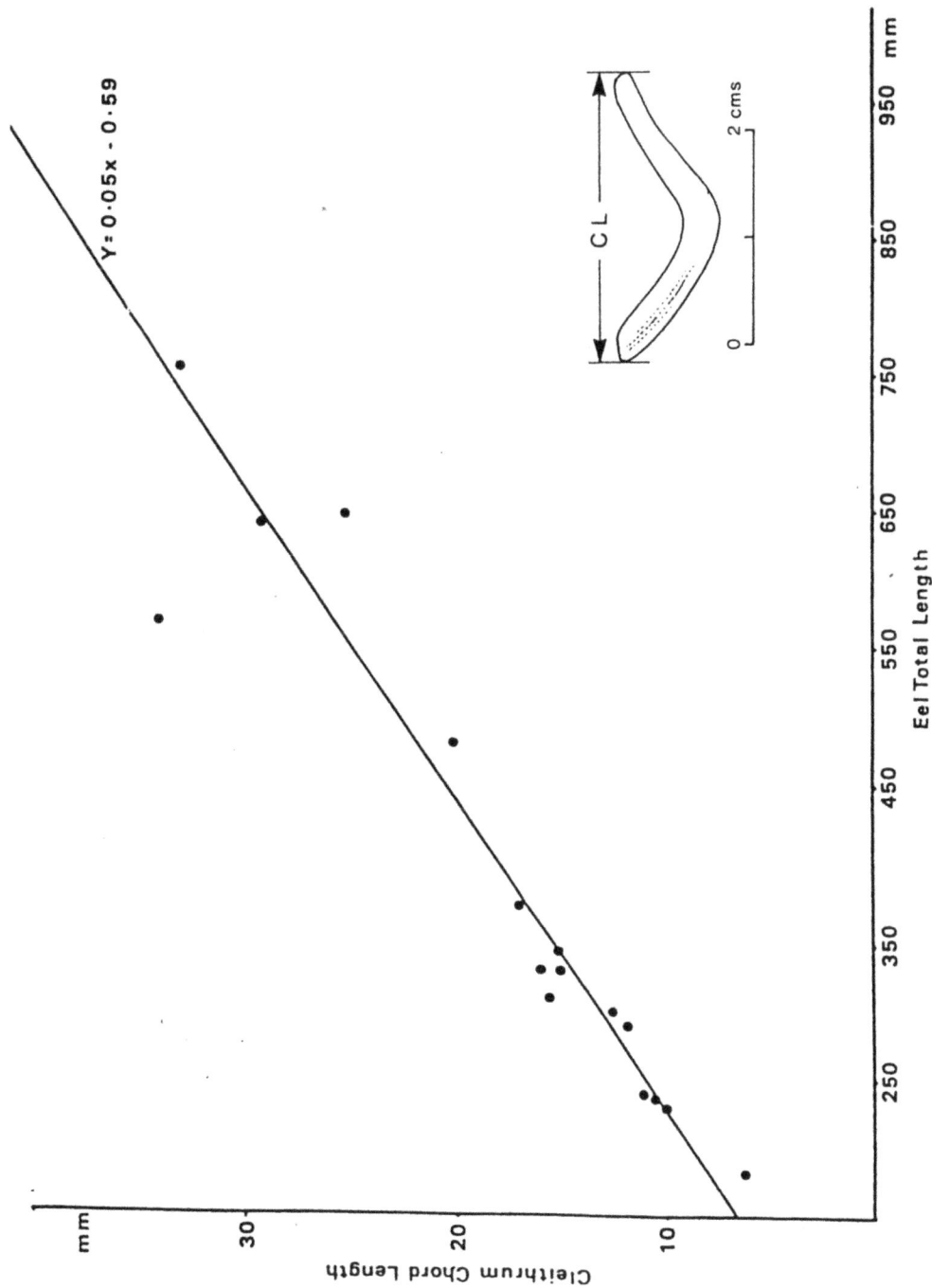

Figure 3. Eel (Anguilla anguilla): relationship of total length of the fish (x axis) to cleithrum chord length (y axis), from a modern sample of 16 specimens. The cleithrum, the most robust bone in the eel, is illustrated.

Acknowledgements

I am grateful to a number of excavators for permission to quote work in typescript before publication and their help in interpretation of the material, especially Mike Morris, Ken Qualmann, and Kevin White. The original bone reports were mostly funded by English Heritage. Difficult identifications were made with the help of Graham Cowles and Alwynne Wheeler and I thank the Trustees of the British Museum (Natural History) for use of the collections on such visits.

REFERENCES

BOURDILLLON, J. (1979) The animal bones of Hamwic: some comparisons. In Archaeozoology (edited by M. Kubasiewicz) Szczecin: Agricultural Academy. 515-524.

BOURDILLON, J. (no date 1) Animal bones from Saxon Southampton; the Six Dials Variability study. Unpublished report No. 4580 to the Ancient Monuments Laboratory July 1984.

BOURDILLON, J. (no date 2) Levels of recording. Unpublished report to the Ancient Monuments Laboratory March 1985.

BOURDILLON, J. (no date 3) The animal bone from Site 32, Southampton: an introduction to the archive from scanning. Unpublished report to the Ancient Monuments Laboratory April 1985.

BOURDILLON, J. (ed) (forthcoming) The environment and economy of Saxon and Medieval Southampton. Southampton Museums.

BOURDILLON, J. & COY, J. (1980) The animal bones. In (ed) P. Holdsworth Excavations at Melbourne Street, Southampton, 1971-76, CBA Research Report 33, 79-121.

COLLEY, S.M. (forthcoming) The Hamwic Pit Project. In J. Bourdillon (ed) forthcoming, above.

COLLEY, S.M., TODD, S.J.P. & CAMPLING, N.R. (1988) Three-dimensional computer graphics for archaeological data exploration: an example from Saxon Southampton. Journal of Archaeological Science 15, 99-106.

COLEMAN, O. (1960) The Brokage Book of Southampton University of Southampton.

COLUMELLA, L.J.M. (1954) Lucius Junius Moderatus Coluella on agriculture (and trees) with a recension of the text and an English translation by E.S. Forster & E.H. Heffner. Volume II. London: Heinemann, Loeb Classical Library.

COTTRELL, P. (forthcoming) Strategies and tactics for sieved recovery. In J. Bourdillon (ed) forthcoming, above.

COY, J.P. (1981) The bird bones. In Excavations in Chalk Lane, Northampton 1975-78 (J. Williams & M. Shaw). Northamptonshire Archaeology, 16, 134 & microfiche 200-211.

COY, J.P. (1983i) Birds as food in prehistoric and historic Wessex. In *Animals and Archaeology: 2 Shell middens, fishes and birds* (eds. C. Grigson & J. Clutton-Brock) BAR International Series 183, 181-195.

COY, J.P. (1983ii) Animal bone In S. M. Davies, Excavations at Christchurch, Dorset, 1981 to 1983 *Proceedings of the Dorset Natural History and Archaeological Society* 105, 43-45.

COY, J.P. (1984) The bird bones. In B. Cunliffe, *Danebury: an Iron Age Hillfort in Hampshire* CBA Research Report No. 52, 526-531.

COY, J.P. (no date 1) Animal bones from medieval and post-medieval phases of Winchester Western Suburbs, Report No. 4910 to the Ancient Monuments Laboratory August 1984.

COY, J.P. (forthcoming) Wild vertebrates, ecology, and diet. In J. Bourdillon (ed), forthcoming, above.

COY, J.P. & HAMILTON DYER, S. (forthcoming) Animal bones from St. Michaels, Southampton: contexts associated with the medieval fish market.

ERNLE, Lord (1961) *English Farming Past and Present* New 6th ed. Heinemann/Frank Cass & Co Ltd with introduction by G E Fussell & O R McGregor.

FINLAYSON, W.D. & BELLHOUSE, D.R. (forthcoming) Empirical studies of sampling designs for the excavation of middens and longhouses on Iroquoian sites in Northeastern North America. Paper contributed to the session on Data Management and Mathematical Methods in Archaeology. World Archaeological Congress, Southampton, September 1986.

GURNEY, J.H. (1921) *Early Annals of Ornithology* London: Witherby.

HARCOURT, R. (1979) The animal bones. In *Gussage All Saints, An Iron Age settlement in Dorset* by G.J. Wainwright. Department of the Environment Archaeological Report No. 10 HMSO, 60-70.

HILLMAN, R. (forthcoming) Animal bones from soil samples at Westgate. In J. Bourdillon (ed), forthcoming, above.

KEENE, D. (1985) *Survey of Medieval Winchester* Winchester Studies Volume 2. Clarendon Press: Oxford.

KEEPAX, C.A. (1981) Avian egg-shell from archaeological sites. *Journal of Archaeological Science*, 81, 315-335.

MALTBY, J.M. (1981) Iron Age, Romano-British and Anglo-Saxon animal husbandry - a review of faunal evidence. In *The Environment of Man: the Iron Age to the Anglo-Saxon Period* (eds) M. Jones & G. Dimbleby BAR British Series 87, 155-203.

MALTBY, J.M. (1985) Patterns in faunal assemblage variability In *Beyond Domestication in Prehistoric Europe*, (eds) G. Barker & C. Gamble. London: Academic Press, 33-74.

NICHOLSON, E.M. (ed) (1929) Gilbert White's Natural History of Selborne, edited by with an introduction and notes by E.M. Nicholson, London: Thornton Butterworth.

PLATT, C. & COLEMAN-SMITH, R. (1975) Excavations in Medieval Southampton Vol. 1, University of Leicester Press.

RIDDELL, W.H. (1943) The domestic goose. Antiquity XVII, 148-155.

SMYTH, J. (1883) Lives of the Berkeleys.

STUDER, P. (ed) (1910) The Oak Book of Southampton Southampton Record Society.

STUDER, P. (1913) The Port Books of Southampton Southampton Record Society.

SOCIAL DIFFERENTIATION FROM ANIMAL BONE STUDIES

F. Gerard Ijzereef

This paper describes the results of work carried out on material excavated by the urban archaeologist of the city of Amsterdam. A paper relating to this study will appear in the papers of the Fifth International Archaeozoological Conference at Bordeaux, 1986 (Ijzereef in press). It uses what seems to me an interesting new approach to zooarchaeological methods, which has influenced my work on other large bone samples. In my work with the Netherlands State Service I find myself in the difficult position of acquiring animal remains for study in large quantities, both from our own service and from urban archaeologists who have no zoological assistance. After the excavation season finishes each year, there is enough archaeozoological work to fill a further five years. Choices therefore have to be made.

In the past large samples were studied, using the standard method of recording a great many data for each individual bone. Here an alternative method was tried, which is suitable in the case of extremely large bone samples, i.e. quantities of 100,000 bones or more, which are of periods of recent date and for which the range of species is already known. The method is appropriate where specific questions are being asked of the material. Another criterion is that the distribution of the bones must be equal over the site and that they are in a similar state of preservation.

The study which is described here did not have the aim of producing a complete list of species but was intended to answer one question of particular interest: what can we learn from animal bones about human diet; and specifically what can animal bones tell us about the social differences between individual households?

The material is from approximately 100 cess pits from two excavations of 17th and 18th century sites. The faunal remains were excavated between 1981 and 1983 at two sites, Waterlooplein and Oostenbergermiddenstraat, by Jan Baart of the Archaeological Service of Amsterdam (Baart et al, 1986). At Waterlooplein two complete blocks of 110 houses were excavated before the construction of a new town hall and opera house (figures 1 and 2). From about 1600 this area was inhabited by Portuguese Jews. At the Oostenburg site, which is located in the harbour area, five houses belonging to the Dutch East Indies Company were excavated: they included five cess pits.

The material was extremely well preserved. It was collected by hand-trowelling, followed by wet-sieving and flotation of samples from the pits. Succeeding layers of fill were seen in most of the cess pits so the samples were divided into smaller fractions or sub-samples, which could be dated to within 25 years. It was clear that the pits must have been cleaned out at regular intervals, because the layers were very thin, and there are usually gaps between datable samples. In many pits the early 17th century fill is overlain by fill from the late 18th century: other pits do not appear to date from before 1700. Most of the houses had one cess pit, but some had more than one and some pits could not be associated with any particular house. In total 165 dated samples were obtained from the cess pits,

Figure 1. Detail from a map of Amsterdam showing the city c. 1650. The square in the centre with four house blocks is the area of the Waterlooplein, the former Vloonburch. The two blocks on the Amstel river were excavated (see Figure 2).

varying in weight between 215g and 47kg.

As well as the cess pits, the material also included ten larger samples from a refuse layer at Oostenburg, dating from between 1595 and 1605, containing 106kg of animal bone. This layer contained refuse from various sources: primary slaughter (mainly skulls with horn cores), discarded food remains, and the refuse from a button-maker (see MacGregor, this volume), a goat-leather worker and cattle phalanges used in children's games.

Although most of the samples are dated to within 25 years, for the purpose of this study they were combined into periods of 50 years.

1. the period around 1600, with material from the non-Jewish layers,

2. the first half of the 17th century,

3. the second half of the 17th century,

4. the first half of the 18th century, and

5. the second half of the 18th century.

METHOD

The bones were separated into nine species or groups of species (see below) and the weights of each group were recorded. In 16% of the material the bone from individual species was both counted and weighed. The bones of the skull, the ribs, the vertebrae, and of the front and hind limbs were weighed separately for cattle, pig, sheep, and goat. The relative frequencies of the weights per taxon or group form the basis of this study. The data were entered on computer and sorted not only by period and location but also by the different percentages of the various species. Thus if the percentage of pig bones is taken to be the indicator for a Jewish or a non-Jewish household, a list can be made of these two household groups. This method is discussed in more detail later.

The material was divided into the following groups:

1. Cattle. The fragments in the cess pits were usually very small; most are of ribs and vertebrae. Skulls and horn cores were not recovered from the cess pits in large proportions, and fairly complete bones occur only in the layer of refuse.

2. Pig. This species was found mainly in the refuse layer and cess pits from Oostenberg.

3. Sheep and goat. Goat horn cores were found only in the refuse layer. All the other bones indicate consumption of mutton at the sites. During the 17th and 18th century the sheep were very large compared with those from earlier periods.

4. Birds. Approximately 90% of the bird bones are from domestic chicken, duck, goose and turkey (<u>Meleagris gallopavo</u>); but many other species also occur.

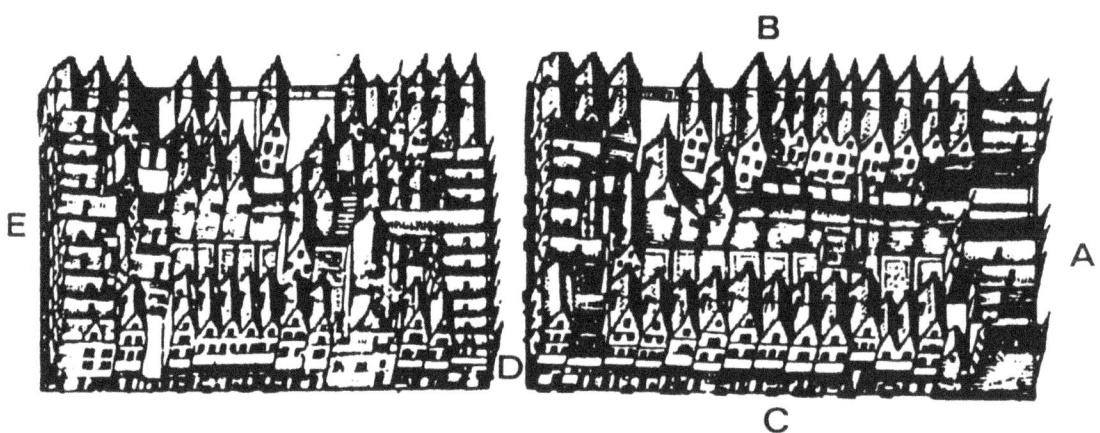

Figure 2. The excavated house blocks. The streets were: A, Zwanenburgerwal
B, Zwanenburgerstraat; C, Lange Houtstraat D, Korte Houtstraat;
E, Leprozengracht/ Waterlooplein.

	cattle weight %	pig weight %	sheep/goat weight %	birds weight %	fish weight %	total weight
Waterlooplein						
ca. 1600	136.1 (81)	10.8 (6)	20.0 (12)	1.2 (1)	0.8 (1)	168.9
17th century	262.6 (74)	7.8 (2)	51.2 (15)	22.0 (6)	9.9 (3)	353.5
18th century	311.9 (85)	2.4 (1)	24.2 (7)	20.3 (6)	10.0 (3)	368.8
1800-1850	4.2 (86)	0.1 (2)	0.3 (6)	0.2 (4)	0.1 (2)	4.9
Total	714.8 (80)	21.1 (5)	95.7 (11)	43.7 (5)	20.8 (2)	896.1
Oostenburg						
18th century	11.2 (52)	4.8 (22)	3.6 (17)	0.4 (2)	1.6 (7)	21.6
Total	726.0 (79)	25.9 (3)	99.3 (11)	44.1 (5)	22.4 (2)	917.7

Table 1. Amsterdam Waterlooplein/Oostenburg: weight (kg) and percentages of the main groups by period.

5. Fish. Fourteen species have been found and preliminary investigations show that 94% of the remains are from marine species. The fish are being studied by F. Laarman (ROB).

6. Rabbit and hare.

7. Molluscs. More than fifty species are present, of which cowrie shells (Cyprea sp.) were found in large quantities at the Oostenburg site. These Pacific species have also been found in American Indian sites, and it is probable that they were transported by the Dutch East Indies Company from Indonesia through Amsterdam to New Amsterdam or New York. Those found at the Waterlooplein are presumably collectors items.

8. Lobster. (Homarus vulgaris).

9. Other remains. Some remains of inedible species or those not normally eaten such as cat, dog, rats and mice were present. Eggshell, feathers, and also coprolites were found.

RESULTS

Almost 1000 kg of animal remains were recovered in the excavations, and from the calculation described above it is estimated that there were over 100,000 individual bones. The clear difference between the layer dating from around 1600 and the later periods when Jewish habitation began can be seen in table 1.

If the results are shown with the cattle omitted from the comparison, a clearer understanding of the relative quantities of the groups with lesser weights can be obtained (figure 3). The weight of sheep bones reaches a peak in the period 1600-1650; birds bones are most common between 1650 and 1700.

A calculation which indicates the numbers of bones involved is shown in table 2. The number of bones has been estimated from counting and weighing a sample of 147.41 kg (16%) of the material. The number of fish bones, c. 45,000, is so great that they clearly need to be the subject of a special study. Chicken bones are also very common: there are more chicken than sheep bones from 1650 to 1700. The peaks for sheep, pig and fish in the period 1750-1800 (figure 3) reflect the inclusion of material from Oostenburg, from the shipyard houses. At that site cattle accounts for 52% of the bones, and sheep and pig for 17% and 22%. The percentage of fish bones here is also very high. In absolute numbers, it means that there are twice as many fish bones as those of any other single species.

Jewish and non-Jewish households

The main criterion used to separate cess pits of the Jewish and non-Jewish households is the percentage of pig bones. On that basis four groups can be distinguished:

1. Zero per cent of pig remains. Jewish households.

2. Zero to one per cent of pig remains. Jewish households. The small quantity of pig remains may represent contamination from later layers.

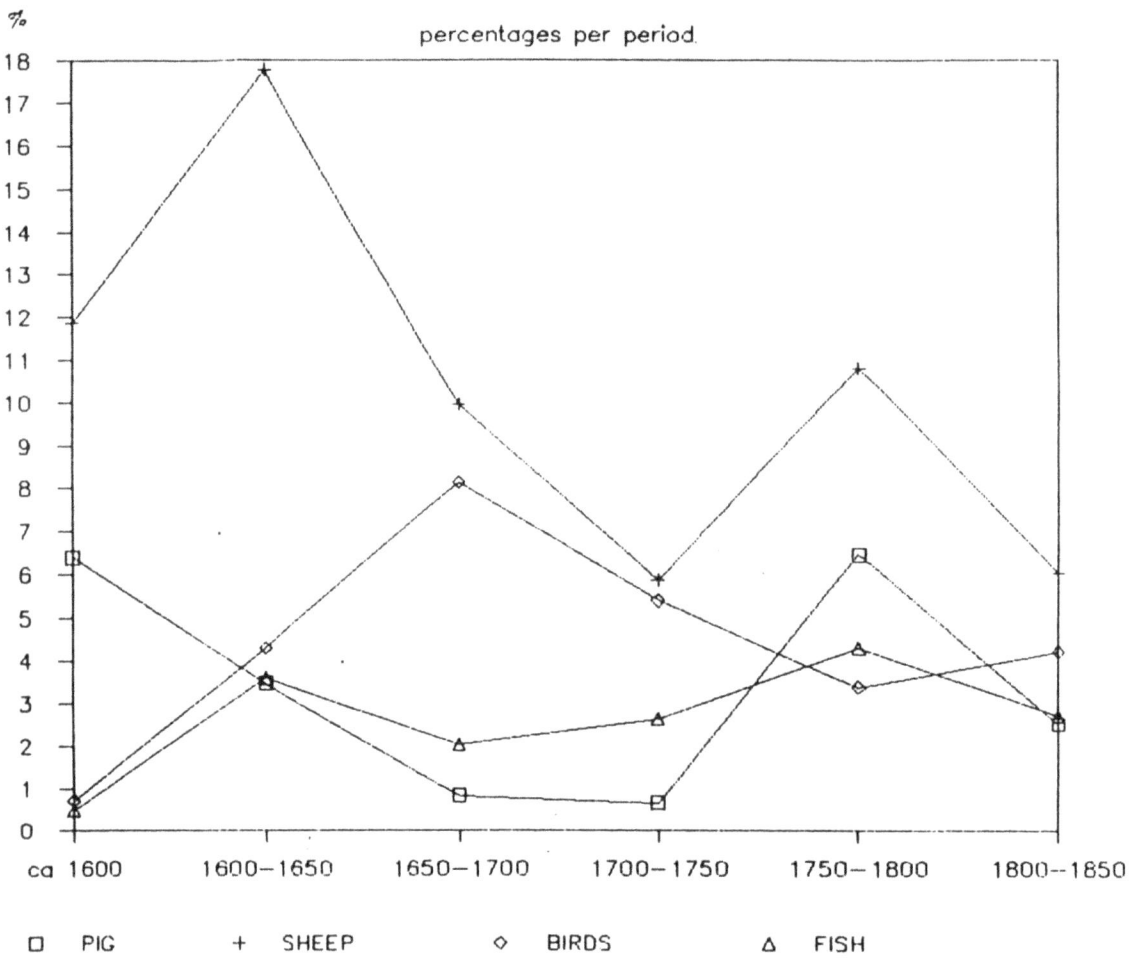

Figure 3. Graph showing percentages of the main species excluding cattle, by period.

	total weight in grams	average weight per fragment	estimated number of bones
Cattle	726,000	28.79	25,217
Pig	25,900	7.06	3,668
Sheep/goat	99,300	7.32	13,565
Birds	44,100	2.66	16,578
Fish	22,400	0.50	44,800
Total	917,700		103,828

Table 2. Numbers of bones from Waterlooplein and Oostenburg. The number has been calculated from counting and weighing a sample of 16% (147.7kg) of the total.

3. One – five per cent of pig remains. Non-kosher households, or Jews with non-Jewish residents living in the same house.

4. More than 5% pig remains (usually about 15%). Non-Jewish households.

There are other very specific differences between Jewish and non-Jewish households. Firstly, apart from the lack of pig remains in the cess pits from the Jewish households there is also an absence of hind limb bones of cattle and sheep. This part of the carcass was considered non-kosher, because of the presence of the sciatic nerve which is difficult to remove (Koen 1970). The material clearly demonstrates that instead of preparing the meat of the hind quarters according to orthdox regulations, there was a preference for selling the hind quarter to non-Jewish butchers. Secondly, there are no calf bones: Jewish butchers bought only adult oxen for kosher butchering. Thirdly: the percentage of chicken bones is very high, whereas in non-Jewish contexts duck is common. The specially butchered and prepared kosher chickens had a lead seal attached to the sesamoid bones in the leg, which showed the day of slaughter. Most of the seals show that the chickens were slaughtered on a Thursday, probably for consumption on a Friday night. One cess pit contained 225 seals; and the chicken bones indicated a MNI of 175 birds. Fourthly the eel (<u>Anguilla anguilla</u>) is absent; it is considered an impure animal. Lower percentages of fish and molluscs complete the picture.

If we look at the distribution of Jewish and non-Jewish households based on these factors, it can be seen that they seem to form a cluster, especially in the left house block. Each of the four groups identified may be examined separately by sorting on the percentages of sheep, birds, fish and cattle. For all households there are criteria other than religious distinctions which indicate social differences. These are based on the amount of money spent on food, personal preference and availability of products in the neighbourhood. Non-Jewish households, for example, would have bought kosher foods. An example of the latter influence may be the absence of evidence for the consumption of veal: calf bones are absent from all houses at Oostenburg, veal was hard to obtain in that quarter. It is obvious that the Jewish group also adapted somewhat to their new habitat. This can be seen from the presence of molluscs, which according to strict orthodox regulations are considered to be unclean. This can also be explained however by the fact that Jews and non-Jews lived in the same house.

If personal preference is discounted, and the prices of the different products are known, it might be possible to deduce the influence of social differences on the variations in meat consumption on the basis of price. Meat prices have been studied by Postumus (1964), and from his study it can be seen that mutton was up to 30% more expensive than beef between 1600 and 1650, but in later periods was always cheaper. Since the highest percentage of sheep bones is found in the sample from 1600 to 1650, this suggests that the area was wealthy at this time. It is the period of great expansion of Amsterdam, and the main phase of the city's Golden Age, which lasted from 1585 to 1670.

In the cess pits in the non-Jewish houses, high percentages of sheep bones correlate with both high numbers of fish bones and the greatest diversity of species, especially freshwater fish. Duck and rabbit are also common. In the cess pits of the Jewish households, higher percentages of sheep correlate with higher percentages of bird bones, of chicken, capon, goose and turkey. Fish are less abundant, but many different species are present.

If the pit samples from 1750-1800 are sorted on the percentage of pig bones (figure 4) it can be seen that they form a series: those from Oostenburg (nos. 5, 9, 20 and 6) have the highest percentages of pig, and also high percentages of sheep and fish. They are followed by two non-Jewish (nos. 250 and 251) and one non-kosher pit (no. 124), and the Jewish cess pits (no. 124) have zero percentage of pig. The percentages of birds, fish and sheep also decline.

Wealthy and poor households

The contents of the cess pits which indicate the wealthiest family (nos. 175, 235 and 36 in table 3) are those from a house on the corner of house block 1, Zwanenburgerwal 35 (figure 2) in the eighteenth century. The percentage of cattle bones is low, and those which were found are from parts of the carcass which are normally indicative of the better quality joints of meat. Hind limb bones are absent as are metapodials, and there are few skull bones. Furthermore, there is a high proportion of chicken, including the 225 referred to above, turkey and goose. A high percentage of fish, from tuna (<u>Thunnus thynnus</u>), herring (<u>Clupea harengus</u>), garfish (<u>Belone belone</u>), and many fresh water species including pike (<u>Esox lucius</u>), perch (<u>Perca fluviatilis</u>), and salmon (<u>Salmo salar</u>) were present, together with high percentages of oyster shell (<u>Ostrea edulis</u>), cowries and other exotic shells. Remains of rabbit, lobster and eggshell were also found.

The faunal assemblage (table 3) which indicates the poorest household, Lange Houtstraat 8, was also found in an 18th century cess pit. It contained 44kg of predominantly cattle and sheep bones, with a surplus of skull, metapodials, and phalanges, all of which were broken open or smashed to obtain the marrow; even the second phalanx was smashed for this reason. This does not point at meat consumption at all, and the composition of the faunal material may indicate that the family lived on the charity of the local butcher. The fish remains indicate a similar situation: skull bones of cod (<u>Gadus morhua</u>) and haddock (<u>Melanogrammus aeglefinus</u>) are the only elements found.

The remaining cess pits form a series between the two referred to above. The percentages of pig, cattle, sheep, birds and fish from a selection of these are also shown in table 3. Zwanenburgerwal 51 was apparently an above average household, judging from the relative proportions of sheep and cattle. Zwanenburgerstr. 25a is a good example of how the faunal remains indicate that the household became poorer in the course of time: the contents of the cess pit suggest that the house was occupied by a rich family in the 17th century and became poorer in the succeeding century. At Lange Houtstraat 38 the pig remains suggest that the initial non-Jewish inhabitants were succeeded by a Jewish household. There is a general overall decline between 1600 and 1750.

The minor percentages of pig at Waterlooplein 26, in the Jewish quarter, are difficult to explain. The material from Oostenburgermiddenstraat 20 contrasts clearly with that from the houses in the Jewish quarter, with its high percentage of pig bones.

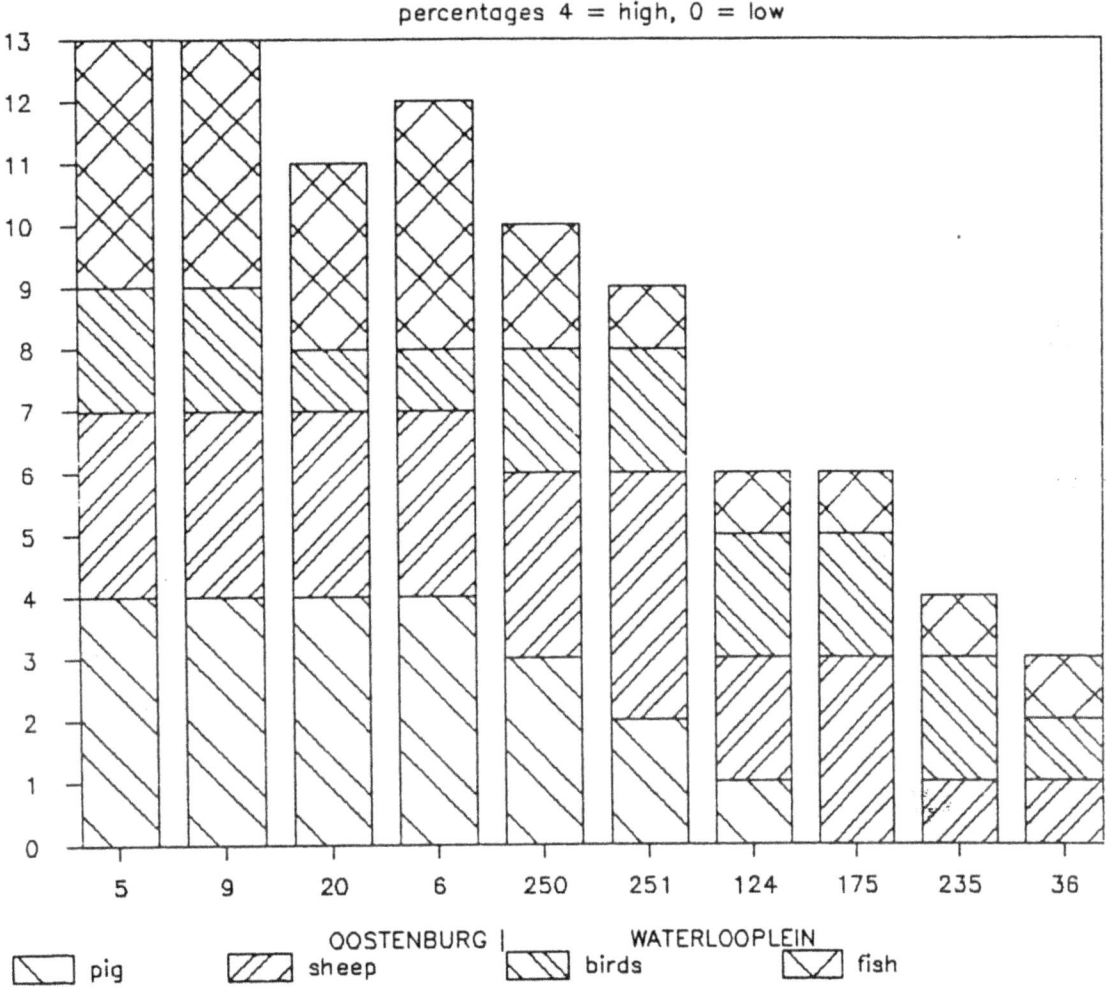

Figure 4. Percentages of pig, sheep, birds and fish from four cess pits at Oostenburg and six from Waterlooplein. The results are arranged by declining percentage of pig.

street + nr	date	pig	cattle	sheep	birds	fish	status
Zwanenburgwal 35	1700-1750	0	68	10	17	5	rich
Zwanenburgwal 51	1650-1700	0	71	16	12	2	average +
	1750-1800	0	75	14	5	7	average +
Zwanenburgerstr 25a	1650-1675	0	54	34	4	9	rich
	1675-1700	0	74	12	12	2	average +
	1700-1750	0	80	15	4	2	average
	1750-1800	0	99	0	0.4	0.1	poor
Lange Houtstraat 8	1700-1750	0.4	93	5	1	1	poor
Lange Houtstraat 38	1600-1650	33	40	8	3	16	average +
	1650-1700	0	84	13	1	2	average
	1700-1750	2	88	4	3	2	average -
	1750-1800	0	85	7	6	3	average
Korte Houtstraat 28	1600-1650	0	77	18	3	2	average
	1700-1725	0.2	85	3	8	4	average
	1725-1750	0	90	5	3	2	poor
	1750-1800	0	84	6	8	3	average
Waterlooplein 26	1600-1650	5	54	31	6	4	average +
	1650-1700	0.8	79	10	8	2	average
	1700-1750	1.7	87	4	3	2	average -
Oostenburgermidden-straat 20	1700-1750	19	51	18	3	10	average +
	1750-1800	18	60	15	1	6	average +

Table 3. Percentages of pig, cattle, sheep, birds and fish bones from eight cess pits. The final column ("status") shows the inferred status of the household.

Cattle remains as an indication of social differentiation

One final analysis was carried out on the samples of all dates which were larger than 2kg. The percentages of weights were sorted into different classes. Cattle were sorted into six classes in bands of 10% varying from class 1 (less than 50%) to class 6 (more than 90%), thus decreasing the percentages of all other species. This division provides a crude method of classification, but permits us to compare the percentages of the other species in each of the groups. The following correlations were found:

1. The group with the lowest cattle percentages always has high percentages of pig and sheep, and usually high proportions of fish and rabbit. Percentages of bird bones are low. We can assume that this group represents the wealthy non-Jewish families.

2. In the second group, with cattle up to 60%, pig may be present or absent, weights of sheep bone are high, and birds and fish are high or average. Rabbit is usually absent. Included in this group are the richer faunal assemblages of both Jewish and non-Jewish households.

3. and 4. These groups form the main bulk of the households of average wealth. Pig is sometimes present in low proportions and I have hypothesised above that this indicates a non-kosher household.

5. The largest group is that where the cattle percentages are between 80% and 90%, and consequently other species are few. This group can be characterised as below average or poor.

6. The final group represents the poor and very poor households.

Although some caution has to be exercised when using this analysis as a basis of comparison it does seem clear that changes in the social composition of the area through time can be illuminated. If the six groups are seen as social groups the following points emerge:

1. Between 1600 and 1650 most households are either rich or above average (figure 5a).

2. Between 1650 and 1700 more of the households are average, and the first group, the rich non-Jewish households, has disappeared (figure 5b).

3. Between 1700 and 1750, the group of poor households appears to have become very large (figure 5c).

4. The period between 1750 and 1800 is characterised by either rich or poor assemblages; the average groups have disappeared (figure 5d). The non-Jewish households from Oostenburg account in principal for the first group. The picture presented does, however, strongly indicate greater social differences in this period than in the others.

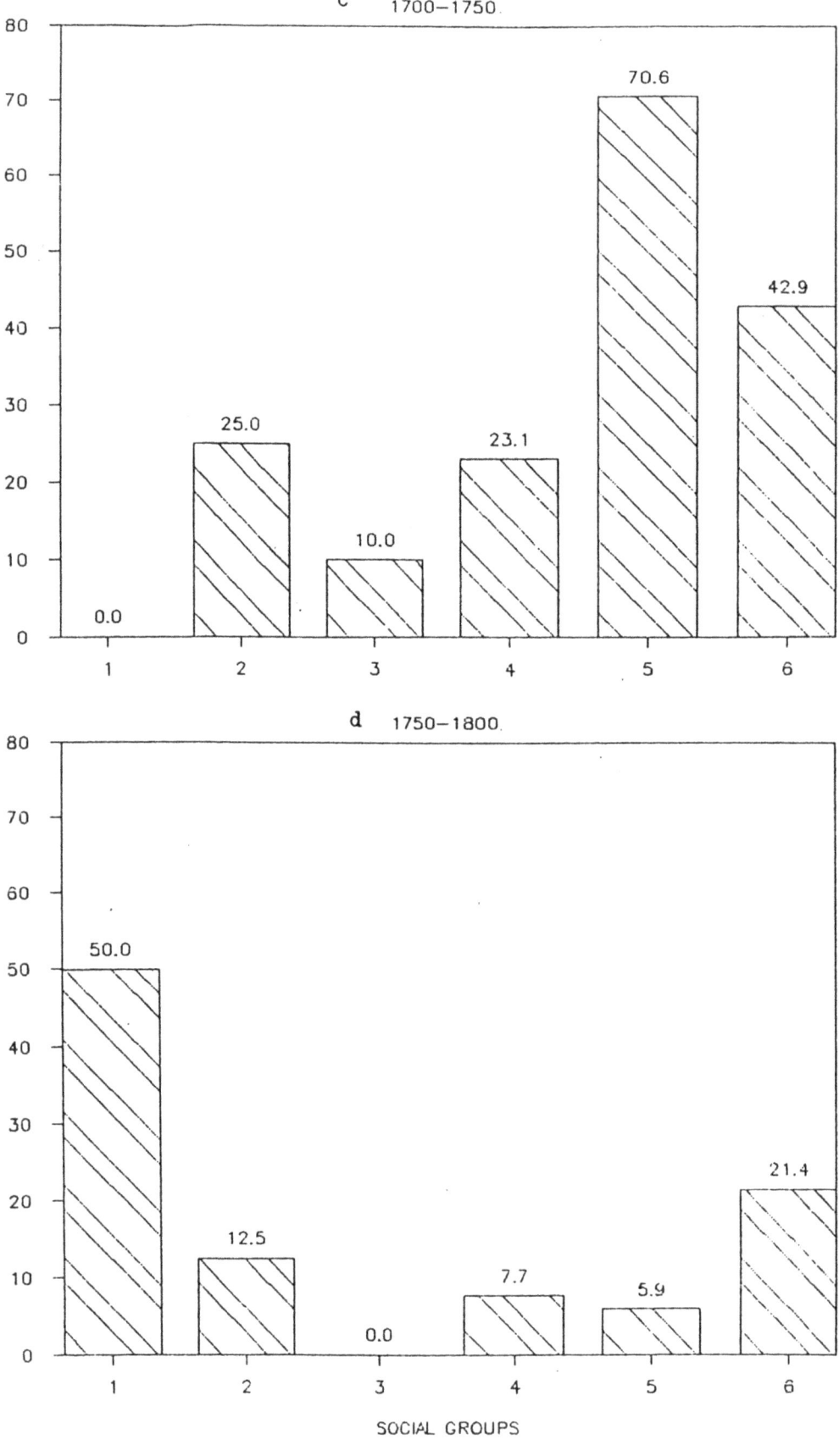

Figure 5. Percentage of cess pits in six classes, grouped according to proportion of cattle bones, for the periods 1600-1650 (a), 1650-1700 (b), 1700-1750 (c), 1750-1800 (d). The cattle bone proportions indicate different social groups, as defined in the text.

Conclusion

The faunal remains from Waterlooplein and Oostenberg provided much more specific information about a household than did the other classes of archaeological finds. The Jewish households could not have been identified in all cases from the artefacts found in the cess pits. Neither the glass nor the pottery indicated minor differences between the households as clearly as did the animal remains. Of course it should not be forgotten that the remains represent only a small part of the refuse which was once thrown away, and minor fluctuations will not always reflect the actual social status of the inhabitants of a house. But this division into six groups has shown that it can provide information on the faunal composition which may well indicate real social differences between households.

What is not in doubt is that this zooarchaeological study has provided new criteria for the analysis of social differences. The house with faunal remains indicating greatest wealth also contained material objects which indicated that it belonged to the wealthiest family. The study demonstrates that it is possible to analyse individual houses by methods other than those which employ generalised data on the economy and on prices. The criteria derived here can now be used in other studies, including excavations of earlier date, to see whether they can serve a purpose as indicators of social stratification.

REFERENCES

BAART, J.M., W. KROOK & A.C. LAGERWEIJ (1986) Excavations in the Oostenburgermiddenstraat (English summary). In: (eds) J.B. Kist et al. Van VOC tot Werkspoor. Utrecht: 219-227.

IJZEREEF, G.F. (in press) Animal Bones and Social Stratification. Proceedings of the Fifth International Archaeozoological Conference. Archaeozoologia 2, in press.

KOEN, B.W. (1970) De vleesvoorziening van de Portugese joden te Amsterdam sinds het begin van de zeventiende ceuw. Jaarboek Amstelodamum 62, 37-48.

POSTUMUS, N.W. (1964) Nederlandse prijsgeschiedenis. Leiden.

THE EFFECTS OF URBANISATION ON HUMAN HEALTH: THE EVIDENCE FROM SKELETAL REMAINS

Tony Waldron

The greater part of this book is taken up with determining the extent to which man's activities have shaped the urban environment, at least so far as the animals he ate and domesticated is concerned. In this chapter I am turning the tables and discussing the ways in which the urban environment might have affected man, and looking to see how far these changes find expression in human skeletal remains. It is a chapter in which potential outstrips hard facts and may be somewhat discouraging on that account.

There is a widespread feeling that, compared with country life, life in the town is somehow 'bad for you'; this feeling is borne out to some extent by mortality statistics which show that the death rates are somewhat higher in towns than in the country as may be seen in table 1. In this table the ratio of age-standardised death rates for different areas of England and Wales are compared with the country as a whole; rates in excess of unity denote higher mortality whilst those less than unity indicate a lower mortality. It will be seen that the ratio increases steadily from rural to urban areas in proportion to the population. Not too much should be read into these data, however, since modern towns attract a disproportionate number of those who are disadvantaged in one way or another and like is not truly being compared with like. Nevertheless, urbanisation was perhaps the greatest social transition which our species has made and various changes in health can be postulated to have followed from it. These include alterations in nutrition, a change in the pattern of disease within the community and changes due to the altered character of work. All of these might in some way affect major demographic variables such as height or life expectation. The skeleton will reflect these changes in lesser or greater degree as will be dicussed below.

Alterations in nutrition

There are two important dimensions of the diet; the protein-calorie intake plus the intake of the macro-elements such as potassium, sodium, calcium and magnesium, and the intake of the essential micro-nutrients such as the vitamins and trace metals. Deficiencies in the first component result, crudely, in starvation and a gross failure in growth, whereas a deficency in the micro-nutrients tends to produce rather characteristic states - the so-called deficiency diseases. Excess of some micro-nutrients may also result in disease. I will deal with height as an index of the first component of the diet later in the chapter and here I will concentrate on the micro-nutrients.

The effects of an imbalance in some of the essential micro-nutrients are shown in table 2. Deficiencies of vitamins B, C and D produce well recognised clinical disorders; beri-beri and pellagra result from a lack of vitamins in the B complex; scurvy is caused by a lack of vitamin C and rickets by a lack of vitamin D. Vitamins A and E do not cause any specific illnesses but an excess of vitamin A may produce a generalised periostitis;

Urban areas with population:	
>100,000	1.03
50,000 - 100,000	1.00
<50,000	0.99
Rural areas	0.95

Table 1: Ratio of urban/rural mortality rates. For each area the age-standardised death rate is compared with the national rate. Ratios in excess of unity indicate a greater than average mortality rate; ratios less than unity indicate a lower than average mortality.

	Deficiency	Excess
Vitamin A	Blindness	Generalised periostitis*
Vitamin D	Rickets;* osteomalacia	Hypercalcaemia; hypercalcinosis
Vitamin E	Prolongs activity of free radicals	
Vitamin B	Beri-beri; pellagra	
Vitamin B_{12}	Pernicious anaemia	None
Vitamin C	Scurvy*	None
Copper	Anaemia; skeletal abnormalities†	[Wilson's disease]
Iodine	Goitre; cretinism	Goitre
Iron	Anaemia*¶	Haemachromatosis
Selenium	Keshan disease; Kashin-Beck disease†	
Zinc	Acrodermatitis enteropathica; skeletal abnormalities†	

* Diseases which can be recognised in the skeleton
† Deficiency states possibly detectable by chemical analysis of bone
¶ Deficiency states probably detectable by chemical analysis of bone.
[Wilson's disease] is associated with high tissue copper levels but these are secondary to the primary defect in copper metabolism.

Table 2: Some essential micro-nutrients and the effects of their imbalance.

this is a condition which is found only in those affluent countries where parents force feed their children with vitamins in the mistaken belief that the average western diet is not sufficently nutritious and so is not something which need concern us here.

Vitamin diseases: Of the vitamin diseases, only scurvy and rickets are capable of pathological diagnosis in skeletal remains and of the two, rickets is by far the more easy to recognise. Scurvy was first described with certainty in the 13th century but there is evidence for its existence in the Old Testament. It has always tended to occur amongst those groups of people who were subjected to a restricted diet. In the Middle Ages it was epidemic in northern Europe during the winter months when fresh fruit and vegetables were unobtainable. The disease became less common in Europe after the potato was introduced and there were severe outbreaks in Ireland and Scotland when the potato crops failed in 1846 and 1847 even though there was no overall deficiency of food.

Scurvy often ended military campaigns and long ocean voyages before the discovery that fresh fruit could prevent it. A deficiency of vitamin C results in abnormal collagen formation which in turn causes capillary bleeding especially beneath the periosteum which leads to the formation of periosteal new bone. The disease has been recognised only rarely in past populations by palaeopathologists (Ortner and Putschar, 1981: 272). Maat (1982) has reported evidence for the disease in Dutch whalers buried at Spitzbergen but his diagnosis is by no means unequivocal. One other feature of scurvy is that the gums become swollen and bleed and teeth may be lost and Wells (1964: 118) has suggested that the loss of teeth associated with severe alveolitis which is commonly seen in medieval skeletons is an example of this condition. There is nothing characteristic about the tooth loss in scurvy, however, and so it is difficult either to refute or confirm this statement although it seems plausible.

Rickets, by contrast, is easy to recognise in the skeleton since it does produce a characteristic picture. It is a disease of children and manifests itself in growing bones, the epiphyses of which are enlarged; the bones of the skull are thinned and in life may feel papery and closure of the fontanelle is delayed. When the child begins to walk or move about, the long bones bend under the combined influence of gravity and muscular tension and the pelvis may become distorted. If vitamin D is re-introduced into the diet normal bone growth will be resumed and the rachitic lesions heal although the bones which have been deformed by the disease remain so. Characteristcally, the long bones develop 'buttresses' in the direction of the main lines of force. Rickets was first described in 1645 by the English physician Daniel Whistler in the doctoral thesis presented to the University of Leyden although the poor would have been familiar with it long before then. It is a disease which was endemic in areas of low incident sunlight and although it is frequently found in medieval skeletons (Wells, 1964: 117) there seems little doubt that its prevalence has increased with urbanisation and particularly with industrialisation. In the great British industrial cities in the 19th century and early 20th century it was very common and children with the stigmata of the disease were seen in great numbers as were adults with bowed legs which were the testimony to their childhood disease. Rickets was a particular hazard to women because pelvic deformities could produce serious obstacles to childbirth and the operation of symphysiotomy (that is, splitting the pubic symphysis) was commonly resorted to in such cases. There are relatively few good dietary sources of vitamin D; the only significant ones are egg yolks, milk products and oily fish such as herring and mackerel but these are inadequate to meet the needs of the body. The majority of vitamin D is formed by the action of sunlight on a precursor substance, 7-dehydrocholesterol, in the skin. The resulting product, cholecalciferol, is usually known as vitamin D3. This is an inactive form and is further metabolised in the liver.

Vitamin D deficiency in adults causes a condition known as osteomalacia in which there is a loss of mineral from the bones, especially the long bones. There is no loss of collagen and this condition differs from osteoporosis, which is common in the elderly, in which there is a loss of both mineral and collagen. In osteomalicia the bones may become deformed and fractures are common and although one might have expected examples from the past, the disease has seldom been seen in palaeopathological material (Ortner and Putschar, 1981: 280).

Trace element imbalance: There are about sixteen trace metals which are considered to be essential for normal growth and development. Deficiencies may produce a specific condition, such as goitre in iodine deficiency or merely a failure to thrive. Our understanding of the relationship between the trace elements and normal health has derived largely from work carried out in animals and relatively few deficiency diseases have been discovered in man. One of the features of the trace elements is that not only do they produce a deficiency state if their intake is too low, but they are also toxic in high doses and so their uptake from the gut is generally subject to elaborate control. The signs of toxity are usually non-specific although some distinctive conditions are associated with toxic levels of iodine, copper and iron (see table 2).

The most important results of trace element deficiency are also shown in table 2. On a world wide scale, the anaemia caused by iron deficiency and the goitre secondary to idoine deficiency are by far the most important. The goitre of iodine deficiency leaves no traces on the skeleton and iodine concentrations in the skeleton are far too low to be used to assess idoine status. Iron deficiency however, can be investigated. It is caused by inadequate intake and by parasites of which the most important are hookworms (<u>Ancylostoma duodenale</u> or <u>Necator americanus</u>). Women are particularly liable to iron deficiency anaemia because of their inevitable loss of blood (and iron) during menstruation and pregnancy; if their diet is inadequate to replace this loss, then their degree of iron deficiency may be profound and children on the breast may also become anaemic because the milk does not contain sufficient iron. There is not the slightest doubt that iron deficiency would have been common in the past but its prevalence is unknown. There are two possibilities for detecting it in past populations, however, from the examination of the skeleton itself, or from the determination of iron levels in the bones.

It has been suggested that iron deficiency in childhood causes changes in the orbits (cribra orbitalia) or in the vault of the skull (porotic hyperostosis). (See, for example, Stuart-Macadam, 1985.) Not all palaeopathologists accept that these changes are pathognomonic of iron deficiency anaemia but at least one study has shown that the iron content of skulls with cribra orbitalia is lower than in those without (Fornaciari et al, 1983). Only 12 skulls were examined in the study and so the results need to be interpreted with caution, but it is at least an interesting pointer for further research.

Bone iron levels are high (over 100 parts per million (ppm)) and, providing one can be certain that no post-mortem uptake has occurred, they could easily be used to assess the iron status of individuals or populations in the past, as has been done by Zainio (1968) on a population of Anasazi Indians in North America none of whom he considered to be iron deficient and hence enjoyed adequate nutrition. Susan Kent (1985), on the other hand, considers that cribra is a condition which follows upon sedentism rather than dietary iron deficiency since a sedentary way of life - such as would occur in towns - is much more conducive to the spread of those intestinal parasites whose presence in the body leads to blood loss. Human parasite ova have been found in abundance in town cess pits (and here one thinks of Andrew Jones and the latrines of York -see, for example, Jones, 1982), but hookworm is not common in the temperate climates; it is, of course, plentiful in the tropics.

Two other deficiency states may briefly be mentioned because they may assume endemic proportions in some parts of the world; zinc deficiency and selenium deficiency.

Zinc deficiency: Zinc deficiency gives rise to a condition in infancy called acrodermatatitis enteropathica which is readily reversible by zinc supplementation. Recently it has been found that zinc deficiency is endemic in young males in some parts of the middle east where it is associated with nutritional dwarfism (Underwood, 1977). Since zinc is relatively abundant in the skeleton - its concentration is typically in the range 10 - 100 ppm, the relationship between zinc status and urban life could be investigated.

Selenium deficiency: This is extremely common in China and parts of asiatic Russia and gives rise to a form of cardiomyopathy known as Keshan disease (after the province in China in which it is found) and Kashin-Beck disease, named after the physicians who described it. Kashin-Beck disease is one of an interesting group of endemic arthropathies (Sokoloff, 1985); selenium levels in the bones are low and so the prospect of being able to attribute joint disease in skeletons to selenium intake is probably not very great at present.

Changes in the pattern of disease

Richard Fiennes, who has made a particular study of the origins of disease, was under no doubt about the effect of urbanisation on the pattern of disease. It is evident (he writes) that the human disease pattern has changed fundamentally since populations became increasingly concentrated in towns and cities. A number of entirely new diseases have appeared, not shared with any kind of animal, which require minimal population densities for survival. Many such diseases must have arisen from pathogens of animal origin, which have become adapted to man as the sole host (Fiennes, 1978). There is not the slightest doubt that the conglomeration of large numbers of people in one place would have increased the risks of contracting both water-borne and air-borne infectious diseases.

As Fiennes and others have shown, it is likely that many common human infections were derived from agents present in domestic animals (see table

Disease	Source	Infective agent
Common cold	Horse	Rhinovirus
Diphtheria	Cattle	Related bacterium
Measles	Dog	Distemper virus
Syphilis	Monkey	Monkey treponemes
Tuberculosis	Cattle	M bovis
Influenza)	Domestic	(
Mumps)	animals;	(Related viruses
Smallpox)	birds	(

Table 3: Origin of some human disease. (Based on data in Fiennes, 1978 and Cockburn, 1963.)

Neolithic

BOVINE TUBERCULOSIS

----\ \-- - - o

ca 1000 AD

HUMAN TUBERCULOSIS

-- - - - - -

ca 150 AD

LEPROSY

13-15th c   18th c

---\ \------------ - - - - - - - - - - - - - o

Figure 1: Relationship between tuberculosis and leprosy. Bovine tuberculosis appeared during the neolithic, to be followed by the disease caused by the human tubercle bacillus in about AD 1000. Bovine tuberculosis disappeared in the 20th century with the wide-scale pasteurisation of milk. Leprosy appeared in Europe ca AD 150 and became less prevalent with the spread of human tuberculosis and disappeared in the indigenous population in the 18th century.

3) and with the advent of agriculture as a way of life, contact between man and animal increased substantially. This would have given the animal viruses and bacteria the opportunity to adapt to their new hosts but the spread of an infection in a community is dependent upon at least two factors, immunity and population density. When an infectiuos agent is brought into contact with a community which has not previously encountered it, the disease it causes spreads rapidly and the case-mortality may be high. Survivors of the initial outbreak will in all probability be immune and subsequent infections will lead to a degree of 'herd' immunity. For an air-borne infection to spread in a community which has already met it requires such large numbers (Cockburn, 1963) that they would only be encountered following the advent of urbanisation. On this account, many of the common infectious diseases are not maintained in primitive societies (Black, 1975). Water-borne infections, can sustain themselves with much smaller populations and for these diseases, the factor which encourages their spread is the pollution of drinking water with sewage, something which was not prevented in Britain until the latter part of the 19th century.

The appearance of a 'new' infectious disease in a population may have altered the natural history of another, older disease. This seems to have happened in relation to tuberculosis and leprosy (figure 1). Humans can become infected with the bacterium which causes tuberculosis in cattle (Mycobacterium bovis) by drinking contaminated milk (or eating contaminated milk products) and they develop a disease which can cause lesions in the bones and joints. Skeletons with the signs of tuberculosis have been found from the neolithic but with continuing association, the bovine mycobacterium evolved into a form which was specific to man (M tuberculosis). This agent is spread through the air and it produces a disease which first afflicts the lungs although it may also spread widely throughout the body and also involve the skeleton. The young women who had consumption in Victorian novels had pulmonary (human) tuberculosis. This form of the disease seems to have appeared in Britain about 1,000 AD.

Bovine tuberculosis co-existed side by side with the new form and disappeared during the present century when milk was pasteurised on a wide scale. It is sometimes thought that only the bovine form of the disease affects the bones but this is not so. For example, in England in the 1940s, when both forms were current, bovine tuberculosis accounted for only 10.8% of all cases of tuberculosis of the bones and joints (Francis, 1958: 91).

The human tubercle bacillus seems to have the capacity to confer some degree of immunity from the bacterium which causes leprosy (M leprae) and it has been suggested that the spread of pulmonary tuberculosis led to the decline of leprosy. The first accounts of leprosy in Europe date from around 150 AD but it was on the decline during the 13th and 15th centuries and the last indigenous case in Britain was recorded in the 18th century.

Leprosy and tuberculosis are relatively easy diseases to diagnose in the intact skeleton and so it is possible to study their occurence in past populations. This does not apply to many of the other diseases shown in table 3, however. Syphilis produces characteristic lesions but none of the others could be diagnosed with certainty and this is true of the great majority of diseases which affect man. Those which leave stigmata on the bones probably account for less than 1% of the total and of those which can be studied in the skeleton, the diseases which affect the joints are much the most common. This means that changes in the pattern of most human disease must be inferred from sources other than the skeleton. Infections with intestinal parasites have already been alluded to and there can be no doubt that their prevalence would have increased greatly in towns because of the poor state of sanitation and the resultant contamination of drinking water. As a result, rather more crude indices of health have to be studied; those which affect demographic factors such as height and longevity.

Height

Height is a good index of the general state of nutrition and factors which seriously disturb the nutritional status will show as a failure to achieve its maximum potential. It is part of folk-lore that people in the past were considerably smaller than they are now (and, by inference, less well nourished). The evidence that this is actually the case, however, is not overwhelming for that part of human history which most concerns us here. Table 4 shows data from various studies in Britain, from those of Pia Bennike in Denmark (Bennike 1985), and from the Interdepartmental Committee on Physical Deterioration which reported in 1904. The mean height of men in Britain was at all times less than it is today but not by a great deal - the maximum difference is shown by the Iron Age population which was, on average, 6 cm shorter than the modern man. However, the range of heights for all the populations is great and all overlap the modern mean. A further factor to take into account when interpreting the data is that the heights of past populations have to be estimated by fitting the lengths of individual long bones into regression equations (such as those of Trotter, 1956, for example); the standard errors in these equations may be of the order of 3-4% which is much the same as the difference in the heights in the British archaeological populations. In Bennike's Danish material there is more variation although from the late Neolithic onwards there is little evidence of any steady trend. The heights of the very early Danish populations are a good deal less than subsequent ones and may truly represent biological differences although the numbers in these samples are never greater than 10. One obvious difference between the Danish and the British data is that modern Danes appear to be at least 5 cm taller on average than their British counterparts, which is surprising.

The Danish data show that male heights during the middle of the 19th and early 20th centuries were lower than at any time since the early neolithic. The same phenomenon occurred also in Britain. In 1904 an Inter-Departmental Committee presented a Report to Parliament detailing its findings in relation to an enquiry into the supposed physical deterioration of the population. This enquiry had been instigated by Parliament because so many volunteers for army service during the Crimean and Boer wars had failed to meet the (not very stringent) standards required. The Committee showed that the average height of young men applying to join the Post Office in 1881 had been 1.67 m - less than the height of men in the neolithic. It rose steadily up to 1903 when the average reached 1.71 m but this was still only the same as during the medieval period. There seems to be little doubt that the reasons for this fall in mean height were connected with industrialisation and urbanisation, the single most important factor being an impoverished diet. Improvements in nutrition during the 20th century have permitted us to achieve and slightly to surpass the height of all but our most remote ancestors.

	Britain		Denmark	
	Male	Female	Male	Female
Mesolithic			1.62	1.54
Early neolithic			1.66	1.52
Middle neolithic			1.69	1.50
Late neolithic	1.72		1.76	1.63
Bronze age			1.72	1.64
Iron age	1.68			
Early Roman IA			1.74	1.62
Late Roman IA			1.77	1.62
Romano-British				
Cirencester	1.69	1.58		
West Tenter St	1.71	1.57		
York	1.70	1.55		
Anglo-Saxon	1.73			
Viking period			1.71	1.57
Medieval	1.71		1.73	1.61
1850			1.65	
1881	1.67			
1886	1.66			
1891	1.67			
1896	1.68			
1901	1.69			
1903	1.71			
1908			1.69	1.59
Present day	1.74	1.61	1.79	1.67

Table 4: Mean height (in metres) of adults at different periods in Britain and Denmark. (British data from various sources; Danish data from Bennike, 1985.)

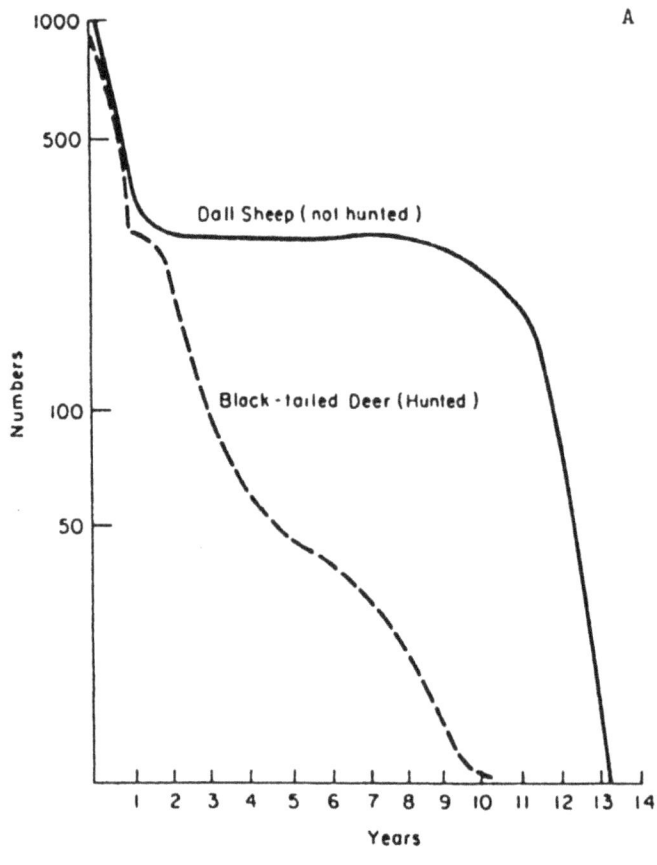
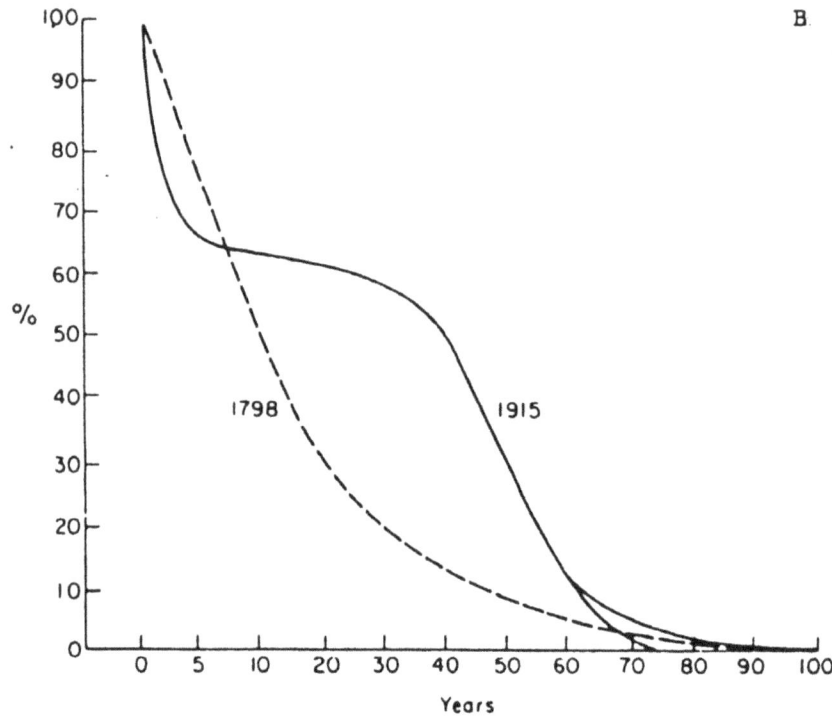

Figure 2A: Survival curves for hunted and unhunted wild animal populations (Black-tailed deer and Dall sheep). This may be compared with the graph for survival of human populations in North Germany (figure 2B.)

Age at death

The grouping together of populations into towns which encouraged the spread of infectious diseases is said to have caused a change in the pattern of mortality which produced a survival curve similar to that of a population of hunted wild animals (see table 2 a & b which are based on Bourliere, 1959). In such a population, there is a more or less steady death rate and relatively few animals reach old age. In a non-hunted wild animal population, by contrast, the death rates are high at both extremes of age but there is a relatively long period when it is steady. It is suggested (Fiennes, 1978:9) that when the prevalence of infectious diseases was high they acted as a human predator but that when the prevalance fell as standards of hygiene improved, the human survival curve altered to approximate to that of an unhunted wild population.

It is possible to derive these kinds of data for the recent past from vital statistics which were collected from about the 18th century onwards but the only sources of material from which to make inferences about life span in more remote times are the skeletons of our ancestors. The methods which are used to age the skeleton are highly subjective and thus prone to much error. It is possible to assign a reasonably accurate age to infants and juveniles on the basis of tooth eruption and epiphyseal fusion but once fusion has taken place it is virtually impossible to do more than place a skeleton within a broad age band; ten year intervals are probably the finest divisions that can be made and many anthropologists and palaeopathologists prefer to place skeletons into descriptive categories - young, mature and old adults, for example. When comparing mean age at death, there is another problem which has to do with the composition of the sample population. The age distribution of skeletons is seldom the same from one site to another; infants and juveniles may be over-represented at one site or under-represented at another (see, for example table 5) and these different age-structures will of course, affect the mean. A population which includes many juveniles will appear to have a lower mean age at death than one which includes many adults although the actual life expectancy of the two populations of which the skeletons are a part may have been the same. But if mean age at death is hard to compute accurately, the life expectation from a group of non-randomly sampled skeletons is virtually impossible to construct. And so the demographic features of past populations have to treated for what they are, best approximations based on inadequate data.

With these caveats in mind, there is a fair body of evidence to suggest that the mean age of death at least, was considerably lower in the past than it is today (table 6). It was between 30 and 40 in men until the beginning of the present century since when it has increased to about 70. In women the mean age of death was always several years younger than in men by contrast with the present day when, of course, women outlive men by several years. The change in the relative length of life in men and women is due to the better obstetric care which has become available during the last 100 years or so (Brothwell, 1972). The overal increase in mean age at death in recent times is almost all due to the containment of infectious diseases which began in the latter part of the 18th century and continued into the 19th, in spite of the worsening conditions which followed in the wake of the industrial revolution. It is difficult to know what were the main causes of death in pre-urban societies and it is probable that we never shall know since we have so few clues from the remains which are available for study.

	Infants (<5)	Juveniles (5-15)	Adults (15+)	n
Cirencester	9.4	7.7	82.8	362
Lankhills	13.0*	16.2†	70.8	284
St Bartholomew's Hospital	11.1	11.1	72.2	18
Trentholme Drive	1.7	6.6	91.7	290
West Tenter Street	6.3	15.2	78.6	112

*Aged <2
†Aged 2-15

Table 5: Age distribution (%) of skeletons from different Romano-British sites. Data from various sources.

	Male	Female
Australopithicenes		
A robustus	19.8	
A africanus	22.9	
Homo erectus	22.8	
Neanderthals	ca 21	
Upper palaeolithic	33.3	28.7
Neolithic	34.3	29.8
Early bronze age (3,000 BC)	33.7	29.5
Middle bronze age (2,000 BC)	36.3	30.8
Late bronze age (1,500 BC)	39.4	32.1
All bronze age	31.3	29.9
Early iron age (1,150 BC)	38.6	31.3
All iron age	31.3	29.9
Romano-British	34.6	31.9
Imperial Rome (120 AD)	40.2	34.6
Dark ages	33.7	31.3
Saxon town	36.0	29.9
Saxon country	34.7	33.1
Medieval town	35.3	30.1
Medieval country	35.3	31.1
1750	40.2	37.3
Present day	70.3	76.7

Table 6: Mean age of death (in years) in adults at different periods. Data from various sources.

Changes due to the altered character of work

Osteoarthritis is the most common abnormality seen in the skeleton and there has been a good deal of discussion on the extent to which this reflects occupational stresses. Calvin Wells was in no doubt that osteoarthritis 'is the most useful of all diseases for reconstructing the life style of early populations' (Wells, 1982: 152). There has been a great deal of epidemiological work to investigate the relationship between work and musculo-skeletal disorders (Lawrence, 1977). Most of these studies, however, have been based on symptoms and few have included radiography as part of the investigation. Reviewing what evidence there is, it seems clear, that in modern populations there is no pattern of osteoarthritis which is so specific to particular occupations that one could state retrospectively the occupation of an individual from changes in the skeleton. Merbs (1983) has undertaken one of the most thorough investigations of a skeletal population, determing the patterns of osteoarthritis in skeletons of an extinct Inuit tribe and relating these to the work practices which had been described by travellers who had visited the Inuit in the past. The bony lesions were then interpreted by the light of the assumed pattern of work.

This methodology is epidemiologically rather unsound in that it makes a posteriori arguments, referring back from (presumed) occupations to skeletal changes. What is required is to establish the pattern of change in those whose work is known and then make a priori inferences, that is by assigning occupations on the basis of finding lesions in a study group which are the same as those in the original reference population. As alraady mentioned, few occupations leave a specific trade-mark on the skeleton and this is a most difficult area of study and Merb's work remains one of the best examples to date of palaeopathological research into occupational arthropathy. It tells us nothing, of course, about the effects of urban occupations.

It seems unlikely that we shall ever be able to tell with certainty what occupation an individual followed on the basis of osteoarthritic change but it is possible to make statements about the patterns of the disease both in individuals and in populations and from this infer something about the stresses which may have determined the expression of the disease. Osteoarthritis is almost certainly a complex of disorders which have a multifactorial aetiology part of which is a genetic predisposition to develop the characteristic changes in the joints. One factor which appears to be important is stress on the joint and although a particular occupation may not lead directly to the development of osteoarthritis it may determine the pattern of expression in those who have the genetic predisposition. If the pattern of joint changes is charted then inter- and intra-group comparisons can be made. If the prevalence of osteoarthritis of the elbow or the hands is significantly greater in one population than another, or if there are distinct patterns within a single population, then it is reasonable to infer that the two groups had different work patterns even though the precise nature of the work will not be known.

A better way of determining occupational stress is by studying the entheses, that is the point at which the muscles insert into the bone. Relatively frequently one finds spurs at these insertions which are thought to represent a high degree of use (and hence of stress on the insertion) of the muscle group concerned. Some individuals - who might be referred to as bone formers - appear to have a greater propensity to develop these spurs

which are also found in association with some of the arthropathies.
Comparing the distribution of these spurs is probably a more reliable index
of work-related musculo-skeketal use than the presence of osteoarthritic
change. To date there have been few studies based of this phenomenon
although Dutour (1986) has looked at a small number of indivudals from two
Neolithic Saharan populations and has claimed to demonstrate different
activities in them.

A difficulty which arises in any study of osteoarthritis or of the
enthesopathies is that the background prevalance in the population is so
great that it may be difficult to determine whether or not there is any
added influence of occupation. To overcome this requires some form of
standardisation which takes into account age and sex differences - the
prevalence of osteoarthritis increases considerably with age, for example -
but the methods of doing this are not within the scope of the present
chapter.

I am actually rather pessimistic about the possibilities of ever being
able to draw definite conclusions about occupations from skeletons even
though intuitively one knows that work does determine pathology - master
masons in the middle ages were much more likely to suffer multiple fractures
or crush fractures to the vertebrae than farmers because their work took
them up scaffolds from which a good number of them would have fallen. It
is a far cry from this, however, to infer that a skeleton with these
injuries must have once belonged to a mason - it could equally have been a
farmer whose cart fell on him. Yet there is an almost irresistible urge to
make this conceptual leap in the dark in the attempt to wring information
from the bones.

Toxic exposures

Urbanisation and industrialisation tend to go hand in hand and the
latter inevitably leads to exposure to potentially hazardous material such
as dusts and metals. We know that silicosis (resulting from exposure to
dusts containing silica) was present in antiquity from the study of lung
tissue from mummies; but the opportunities to study diseases of the soft
tissues are rare (in Britain almost to vanishing point) and so inferences
about exposure to toxic materials must be based on the examination of bones
and teeth. With almost no exceptions, the toxic materials which would have
been encountered in the past leave no characteristic diseases on the hard
tissues; fluoride is an exception and fluorosis has been documented in
skekletons from areas where toxic levels of fluoride occur naturally in the
water (Ortner and Putschar, 1981:288). The assessment of exposure can in
theory be based on measurements of the concentration of the particular
element in the bones or teeth. In practice, however, few toxic elements
lend themselves to this study because their concentration in the hard
tissues is not a function of the amount in the body as a whole (the total
body burden); that is to say, they are not bone-seeking elements. Of
those which are, however, lead is perhaps the most important.

Lead was first used as a source of silver since its principal ore,
galena (lead sulphide) contains as an 'impurity' a certain amount of silver.
In due course lead came to be used for its own sake and it was soon
established as a most useful metal. It is easy to extract from the ore, it
has a low melting point so that it can be cast without difficulty, it is
malleable and easily worked, it can be joined together using nothing more

than a red hot iron and it resists corrosion. In short, lead is a perfect metal for societies with a relatively primitive technology except that it is also highly toxic.

As is well known, the Romans were the first major users of lead and it can safely be assumed that they poisoned themselves to a considerable degree with it for not only did they use it in various forms of construction but they also contaminated their food and drink with it to a high degree (for further details see Waldron, 1973). With the building boom in the middle ages, lead also assumed considerable importance and all manner of buildings from the great cathedrals and abbeys to those of much humbler construction were lapped in lead. Matthew Paris describes how William of Trumpington extended some of the buildings of St Albans Abbey 'all of which were covered with lead, not without considerable expense' (Vaughan, 1986:48).

Exposure to lead can be determined from bone (or tooth) lead concentrations since well over 90% of the total body burden is in the skeletal tissues. Care has to be taken when taking samples for analysis, however, since the concentration of lead in the bones and teeth is not constant; hence bone for anlysis should always be taken from the same site and, if teeth are to be analysed, then the same type of tooth must be used throughout. Care must also be taken to overcome the problem of post-mortem contamination. This is most extreme when samples are taken from a skeleton from within a lead coffin when lead concentrations of between 1-2% may be found (the normal concentration would be two to three orders of magnitude less than this). A number of ways of overcoming this and other problems have been suggested (Waldron, 1988) so that reliable data can be obtained.

For the basis of making comparisons between populations, it is the relative amount of lead in the skeleton which is important since the absolute values are bound to vary from laboratory to laboratory depending on analytical techniques, bone or tooth used, and so on. However, the results from one laboratory can be compared using a non-exposed population as the standard. In my own series, bones from a neolithic population have been used as the standard and results from other populations expressed as a ratio of the 'natural' exposure. The results for several periods are shown in table 7. It can be seen that exposure increased steadily until the 18th/19th century when it began to fall again. The very high exposure during the medieval period was due principally to contamination of food and drink either deliberately - as with the adulteration of wine with lead to improve a poor (unsaleable) vintage - or accidentally from the use of poorly fired lead glazes or cheap pewter. The adulteration of food and drink continued well into the 19th century and the beginning of the 20th but in recent years, the amount of lead in the diet has decreased greatly and more than compensated for the increased amounts of lead in the air; thus the present day urban population is probably less exposed to lead than any other for 2,000 years.

The way forward

In many places in this book, reference is made to the need for the specialist to be involved in the work of recovery; I make no apology for reiterating that plea here on behalf of the examiners of human skeletons. It is axiomatic that the quantity of information derived from the skeleton is in inverse proportion to the amount recovered. Poorly preserved, badly damaged or very incomplete skeletons will seldom repay the efforts taken to

Neolithic	1
Iron age	3.5
Romano-British	7.0
Medieval	13.0
18th/19th century*	10
Present day†	4

*Estimated from documentary sources
†Based on analyses of rib samples obtained at autopsy

Table 7: Relative degrees of lead exposure at different periods in Britain.

remove and examine them. The classification of diseases generally requires that the distribution of pathological changes is established throughout the skeleton; if the hands and feet are missing, for example, it may be impossible to classify the arthropathies which are the most common of the diseases seen in the skeleton (Rogers et al, 1987). Much wasted effort can be saved if the bone specialist is given the chance to see recovery in progress; he or she can suggest which skeletons are worth removing and which are not; an examination in situ may salvage some information from a friable skeleton which disintegrates during its later passage from soil to bench; and some rudimentary anatomical instruction can be given to the diggers. It is helpful to know that you should expect to find eight carpal bones at the ends of the bones of the forearm, for example. Sieving recovers small human bones, such as the phalanges, as well as small animal bones and excavators sometimes need reminding of this.

In towns where the population below ground may be substantially the same as that above ground, the strategies for recovery must be designed to be as cost-effective as possible. State of preservation is one important criterion attention to complete recovery is another. A third may be secure dating.

One of the purposes of the study of human remains is to establish temporal trends in demographic or pathological features of past populations and to attempt to relate them to social or environmental changes occurring at the same time. A typical grave-yard may be used over several centuries and if it has been greatly disturbed, the temporal sequence may be difficult to establish; skeletons in coffins with burial plates in situ are, unfortunately, all too rare. The information which can be obtained from 200 skeletons spanning three or four centuries may be less than could be gained from a smaller number of skeletons from a closely dated feature such as a plague pit.

Numbers cannot be overlooked either since inferences drawn from small numbers are seldom epidemiologically reliable. Human bone specialists always deal with a non-random sample which reflects the characteristics of the population from which it derives to a lesser or greater degree which cannot be determined. With large samples some statistical imperfections may be smoothed out. By all means look at small number of skeletons for the earliest case of rheumatoid arthritis may be amongst them, but reports on small samples will say nothing useful about populations. In urban sites it is much bettter to dig one large cemetery carefully and completely than a dozen small ones (and small means less than a hundred) unless there is the opportunity to amalgamate data or there are compelling reasons. The bones may come from a leper hospital, for example, or be closely dated like the Mary Rose, or they may be the only bones from the town or period - one would not disdain a handful of mesolithic skeletons. As a general rule, however, epidemiological work requires samples of several hundreds otherwise, as soon as the sample is sub-divided into age and sex classes the numbers in each sub-division rapidly become too small to permit of valid observations.

Finally, soil samples should always be taken if trace element analysis is to be considered, preferably from the abdominal region as they can be used also to look for parasite ova.

As I said at the beginning of this chapter, circumstances would compel me to concentrate on what might be learned about the effects of town life on man, rather that what has actually been learned. It will have been seen, I hope, that there is plenty of potential but to achieve it, calls for a high degree of collaboration between those who recover the bones and those who examine them.

REFERENCES

P. BENNIKE, (1985) Palaeopathology of Danish Skeletons, Copenhagen, Akademisk Forlag.

F.L. BLACK (1975) Infectious diseases in primitive societies, Science, 187, 515-518.

F. BOURLIERE (1959) Lifespans of wild animals, In: The Lifespan of Animals, London, Ciba Foundation, 91-92.

D. BROTHWELL (1972) Palaeodemography and earlier British populations, World Archaeology, 4, 75-87.

A. COCKBURN (1963) The evolution and eradication of infectious disease, Baltimore, Johns Hopkins Press.

O. DUTOUR (1986) Enthesopathies (lesions of muscular insertions) as indicators of the activities of neolithic Saharan populations, American Journal of Physical Anthropology, 71, 221-224.

R.N. T-W-FIENNES (1978) Zoonoses and the origin and ecology of human disease, London, Academic Press.

G. FORNACIARI, F. MALLEGNI, D. BERTINI and V. NUTI (1983) Cribra orbitalis and elemental iron in the Punics of Carthage, Ossa, 8, 63-77.

J. FRANCIS (1958) Tuberculosis in animals and man, London, Cassell.

INTER-DEPARTMENTAL COMMITTEE ON PHYSICAL DETERIORATION, (1904) Report, Vol 1 - Report and Appendix, Cd 2175, London, HMSO.

A.K.G. JONES (1982) Recent finds of parasitic ova at York, In: Proceedings of the 4th European meeting of the Paleopathology Association, Ed. G.T. Haneveld and W. R. K. Perizonius, Middlebourg, PPA, 229-233.

S. KENT (1986) The influence of sedentism and aggregation on porotic hyperostosis and anaemia: a case study, Man, 21, 605-636.

J.S. LAWRENCE (1977) Rheumatism in Populations, London, Heineman Medical Books.

G.J.R. MAAT (1982) Scurvy in Dutch whalers buried at Spitzbergen, In: Proceedings of the 4th European meeting of the Palaeopathology Association, Ed G.T. Haneveld and W.R.K. Perizonius, Middlebourg, PPA, 82-93.

C.F. MERBS (1983) Patterns of activity induced pathology in a Canadian Inuit population, Ottawa, National Museums of Canada.

D.J. ORTNER and W.G.J. PUTSCHAR (1981) Identification of pathological conditions in human skeletal remain, Washington, Smithsonian Institution Press.

J. ROGERS, T. WALDRON, P. DIEPPE and L. WATT (1987) Arthropathies in palaeopathology: the basis of classification according to most probable cause, Journal of Archaeological Science, 14, 179-193.

L. SOKOLOFF (1985) Endemic forms of osteoartritis, Clinics in the Rheumatic Diseases, 11, 187-202.

P. STUART-MACADAM (1985) Porotic hyperostosis representive of a childhood condition, American Journal of Physical Anthropology, 66, 391-398.

M. TROTTER (1956) A re-evaluation of stature based on measurements of stature taken during life and of long bones after death, American Journal of Physical Anthropology, 9, 79-125.

E.J. UNDERWOOD (1977) Trace elements in human and animal nutrition, 4th edition, London, Academic Press.

R. VAUGHAN (1986) Chronicles of Matthew Paris, Gloucester, Alan Sutton.

H.A. WALDRON (1973) Lead poisoning in the ancient world, Medical History, 17, 392-399.

T. WALDRON (1988) The heavy metal burden in ancient societies, In: Trace elements in environmental history, Ed G. Grupe and B. Herrmann, Berlin, Springer-Verlag, 125-133.

C. WELLS (1964) Bones, bodies and disease, London, Thames and Hudson.

C. WELLS (1982) The human burials, In: Romano-British cemeteries at Cirencester, Ed A. McWhirr, L. Viner and C. Wells, Cirencester, Cirencester Excavation Committee, pp 135-202.

L.G. WILSON (1975) The clinical definition of scurvy and the discovery of vitamin C, Journal of the History of Medicine, 30, 40-60.

E.C. ZAINIO (1968) Elemental bone iron in the Anasazi Indians, American Journal of Physical Anthropology, 29, 433-436.

URBAN-RURAL VARIATIONS IN THE BUTCHERING OF CATTLE IN ROMANO-BRITISH HAMPSHIRE

Mark Maltby

INTRODUCTION

This paper will discuss some of the butchery marks observed on cattle bones recovered during recent excavations of Romano-British sites in Hampshire. Particular emphasis will be placed upon the results of butchery analysis of samples from Winchester, since this is one of several major Roman towns in southern England that have produced evidence for the large scale processing of cattle carcases. Characteristically this evidence consists of substantial accumulations of cattle bones dominated by particular skeletal elements (Maltby 1984a). The presence of specialist butchers and possibly abbatoirs in such towns is to be expected in order that an adequate supply of meat was provided efficiently for the inhabitants. Apart from meat, the distribution of other animal products such as marrow, horns, skins and raw material for boneworking is likely to have depended heavily upon the presence of butchers or similar specialists resident in the towns.

The two major Roman towns in Hampshire, Winchester and Silchester, have already provided evidence for such activities. Indeed, late nineteenth century excavations at Silchester produced large accumulations of cattle horn cores (Jones 1892). Recent excavations of the Silchester defences provided evidence for the dumping of cattle head and foot bones in some numbers in an area near the South Gate during the late 1st Century AD (Maltby 1984a; 1984b).

Archaeologists in Winchester have also recovered evidence for the disposal of large quantities of particular types of butchery and boneworking waste. Excavation of an early Roman ditch in the Western Suburbs of the Roman town revealed large quantities of boneworking waste dominated by cattle scapula fragments and, to a lesser extent, radius, tibia and metapodial fragments of both cattle and horse (Coy & Bradfield n.d.).

Excavations in the Northern Suburbs of Winchester have produced other early Roman ditch deposits that contained large numbers of split major meat-bearing limb bones of cattle (humerus, radius, femur, tibia) (Pfeiffer n.d.). Later Roman deposits in the same area have provided further evidence for the dumping of large quantities of cattle carcase waste. Several of these dumps were also heavily biased towards the upper limb bones. One pit included a concentration of horn cores belonging to at least 24 cattle (Maltby n.d.1). There is evidence therefore that the area was at times used for the disposal of waste from large scale butchery, hornworking and boneworking activities throughout much of the Romano-British period.

It is fortuitous that these and other faunal assemblages from Winchester (Maltby n.d.2) and Silchester (Grant 1985) do not stand in isolation. Excavations of a variety of rural settlements over the last 20 years have provided several samples which can be directly compared with those from the urban and suburban sites. The largest of these rural assemblages was recovered from the Iron Age and Romano-British settlement at Owslebury (Collis 1970). From there nearly 82,000 animal bone fragments of 1st - 4th century AD date have been examined (Maltby n.d.3). Several samples from other chalkland settlements of late Iron Age and early Romano-British date have also been studied. These include the enclosure systems at Winnal, Down/Easton Lane (Maltby 1985a; n.d.4) two kilometres to the north east of Winchester; the banjo enclosure in Micheldever Wood (Coy 1987) further to the north of Winchester; Abbotstone Down (Maltby n.d.5), near New Arlesford; and Cowdery's Down (Maltby 1983) and Brighton Hill South (Maltby n.d.6) near Basingstoke. Further west, Romano-British samples have been obtained from Old Down Farm (Maltby 1981) and Balksbury (Maltby n.d.7), near Andover and Little Somborne (Maltby 1984c) in the Test Valley to the south east of Stockbridge.

On the south coast, excavations at the fort of Portchester Castle produced a large assemblage of later Roman date (Grant 1975). Samples have also been examined from the small town at Neatham, near Alton in north-east Hampshire (Done 1986). These have provided data from other types of settlement for comparison.

Although this survey will tabulate the frequency of observations of various types of butchery marks encountered in the various samples, the figures should be treated with caution. For example, there will inevitably be more observations of butchery on distal portions of humeri than at the proximal ends, because more distal halves will have survived the destructive effects of animal gnawing and weathering. They do, however, provide a guide to the relative frequency of butchery marks encountered.

SELECTED ROMAMNO-BRITISH SAMPLES FROM HAMPSHIRE

This section will describe and discuss the types of butchery marks observed on the major skeletal elements of cattle in the 13 samples examined by the author listed in Table 1. Most of the evidence for urban butchery used in this survey is taken from the samples from the later Roman deposits discovered in the excavations in the Northern Suburbs of Winchester (Maltby n.d.1). It will be shown that although parallels of the butchery techniques have been found in other parts of Winchester and in other Roman urban settlements, most of them have been encountered only rarely, if at all, on the Late Iron Age and Romano-British rural sites recently investigated. Not every sample from Winchester or Silchester has produced evidence for all these techniques, and, consequently the butchery marks described in this paper should not be regarded as typical of all Romano-British urban samples. Furthermore the full range of butchery variability may not as yet have been encountered. The main aim is to describe some of the techniques employed, in order to provide a basis for futher research into their origins, distribution and frequency in Roman Britain.

The major rural assemblage used in the comparisons is from Owslebury, from where the data are divided into three phases spanning the Late Iron Age and Romano-British period. Inevitably the butchery observations from the smaller samples were often too limited in number to add much information and they will be used mainly to supplement the data from Winchester Northern Suburbs and Owslebury.

Site		Date	Butchered Fragments	Total Cattle	% Butchered
Winchester,	Northern Subs.	3-4 A.D.	1431	4364^	33
	Staple Gardens	Roman	200	580	34
Silchester,	South Gate	Roman	83	585	14
	Manor Farm	Roman	12	106	11
Owslebury		1st A.D.	433	4155	10
		1-2 A.D.	153	1783	9
		3-4 A.D.	149	3759	4
Winnall Down/Easton Lane		LIA-ER	57	824	7
Cowdery's Down		LIA-ER	26	253	10
Abbotstone Down		LIA-ER	24	264	9
Brighton Hill South		LIA-ER	27*	536	5
Little Somborne		Roman	14	243	6
Balksbury, 1973		Roman	23	333	7

^ = total fragments excluding loose teeth, articulated bones and fragments from sieved samples
* = excludes sawn horn cores
LIA-ER = late Iron Age-early Romano-British

Individual butchery observations were recorded using the Ancient Monuments Laboratory's system of coding. They are stored in site animal bone archives at the Faunal Remains Unit, University of Southampton. Summaries of butchery data used in this survey can be found in the following reports:-

Winchester, Northern Suburbs	-	Maltby (n.d.1)
Staple Gardens	-	Maltby (n.d.2)
Silchester, South Gate	-	Maltby (1984b)
Manor Farm	-	Maltby (1984b)
Owslebury	-	Maltby (n.d.3)
Winnall Down/Easton Lane	-	Maltby (1985a/n.d.4)
Cowdery's Down	-	Maltby (1983)
Abbotstone Down	-	Maltby (n.d.5)
Brighton Hill South	-	Maltby (n.d.6)
Little Somborne	-	Maltby (1984c)
Balksbury, 1973	-	Maltby (n.d.7)

Table 1. Cattle Fragments with butchery marks from late Iron Age and Romano-British assemblages from Hampshire.

Quantification of butchery is fraught with methodological problems (Maltby 1985b). Comparisons between Owslebury and Winchester, for example, are made difficult by factors of differential preservation. Fewer observations of butchery were made in the former sample because the assemblages were more heavily eroded. The other rural sites in this analysis have also consistently provided a low proportion of butchery marks (Table 1). Such figures to some extent reflect the greater intensity of cattle butchery in Winchester and Silchester, particularly in the degree of chopping associated with dismemberment, marrow extraction and boneworking. However, they are also biased by the poorer preservation of bones from the rural sites resulting in the greater destruction of butchery marks.

Mandible

Butchery marks were divided into 15 categories according to their location, the implement used and the depth and direction of the cuts or chops. The categories and abbreviations used are defined in Table 2. Several specimens bore more than one type of butchery mark. In the late Roman deposits at Winchester Northern Suburbs, the most common types were chop marks located on the caudal portion of the ramus. These were usually situated near the condylar process and were made when the mandible was detached from the skull. Most frequently these chop or saw marks were located on the caudal surface of the mandible, as illustrated in Plate 1 (Type J12). This specimen in fact bore deeper marks than most examples and had been completely chopped through towards the ventral side. In other cases superficial chop marks were located on the lateral aspect (and very occasionally on the medial) of the ramus, usually just below the condylar process (Type J11). In only two cases were knife cuts observed in this area (Type J8). By contrast, such knife cuts occurred more frequently than chop marks at Owslebury in the early Romano-British contexts. Unfortunately, insufficient mandibles were recovered from 3rd-4th century features, to indicate whether chop marks became more common in the period contemporary with the assemblages from Winchester Northern Suburbs. Both knife cuts and superficial chop marks were found on the lateral aspects of the rami of mandibles in Late Iron Age - Early Romano-British levels at Winnall Down, Cowdery's Down, Brighton Hill South and Balksbury, whilst only knife cuts were observed in the sample from Abbotstone Down.

Although these marks were probably all associated with the detachment of the mandibles from the skull (corresponding marks have been found on the zygomatic and temporal bones at Winchester and Owslebury), variations in the implements used and in the exact location of the butchery marks do provide evidence for subtle differences in the techniques employed. Only in the Winchester Northern Suburbs sample were blows directed mainly on to the caudal aspect of the ramus. Chops on the lateral aspect were more common in the other urban samples from Winchester and Silchester. Such variations in the direction of the chopping may have depended upon the position in which the head lay when the blow was delivered. What is clear is that whereas the use of knives to separate the mandibles from the skull was a technique commonly employed in the Iron Age and into the Romano-British period at some rural settlements, it was largely replaced by the use of heavier cleavers by butchers in Winchester and Silchester. Both types of implements appear to have been used on the later Romano-British rural sites.

There were more consistent contrasts between urban and rural samples in the butchery marks located on the diastema (the area between the incisors and cheek teeth). At Owslebury, knife cuts were commonly encountered on the lateral (buccal) aspect running mainly in a dorso-ventral direction. Similar marks were observed in all the other rural samples (Type J1). Only one such specimen was noted in the large sample from Winchester Northern Suburbs. Instead, superficial chop marks running in a dorso-ventral direction were encountered on both the medial (lingual) and lateral surfaces (Types J5-J6). In addition, at least seven specimens bore chop or saw marks on the medial and ventral surfaces made during the splitting of the mandibles through the symphyseal surface (Type J7). A similar specimen was found at Cowdery's Down but only two observations of superficial chop marks on the diastema were made at Owslebury (Table 2). Knife cuts were encountered on the diastema of mandibles from Staple Gardens, Winchester and at Silchester but only on the medial surface (Type J2).

Again a number of reasons may account for the variability in the butchery marks on the diastema. Most of the marks were probably associated with the separation of the mandibles and/or the removal of the tongue. Knife cuts on the lateral aspect below the cheek teeth (Type J14) were found at Owslebury, Winnall Down and Cowdery's Down. These were probably made during the removal of cheek meat. No such marks were found in the samples from Winchester and Silchester.

Scapula

Thirteen types of butchery marks were recorded (Table 3). As in the case of the mandibles, very few of the marks on scapulae from the urban sites were made with knives, whereas knife cuts were frequently encountered at Owslebury and most of the rural settlements. The main contrast concerned the methods used to disarticulate the scapula from the humerus. At both Winchester and Silchester it was common for the glenoid to have been completely chopped through in an oblique/axial - medio-lateral direction (Type S1). Usually this resulted in the removal of the superglenoid tuberosity and the cranial part of the glenoid from the rest of the scapula. In some specimens, however, the blow was delivered (usually from the lateral side) closer to the caudal border of the articulation. In a few instances at Winchester Northern Suburbs the articulation had been chopped through on both the cranial and caudal sides (Type S2). Only one close parallel was recorded from a late Roman deposit at Owslebury, although four specimens of Iron Age date from those excavations had been butchered in a similar manner. At Owslebury, Winnall Down and Brighton Hill South knife cuts running horizontally along either the medial or lateral aspects of the distal scapula (Type S8) showed that different implements and techniques were used for the separation of the forelimb at this point. Occasionally such marks around the glenoid appear to have been made with a heavier implement (Type S3).

The most frequently observed butchery marks encountered in the samples were those associated with the removal of all or part of the lateral spine of the scapula during filleting. Implements used for this varied, and marks made with cleavers, heavy-bladed knives and occasionally saws or serrated knifes were found (Type S4). The actual damage to the spine varied from superficial chops mainly situated near the neck of the scapula, where the spine rises sharply, to more severe butchery where all of the projecting spine had been removed. Plate 2 illustrates an example from Owslebury where most of the spine towards the distal end has been removed and the lateral border of much of the rest has been trimmed off with the

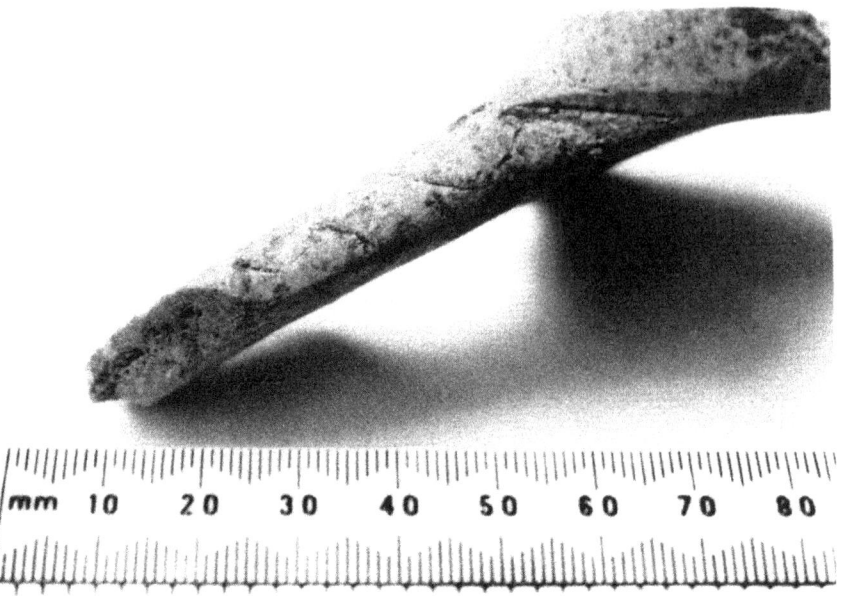

Plate 1 - Cattle mandible from Winchester Northern Suburbs showing chop marks on the caudal aspect of the ramus (Type J12).

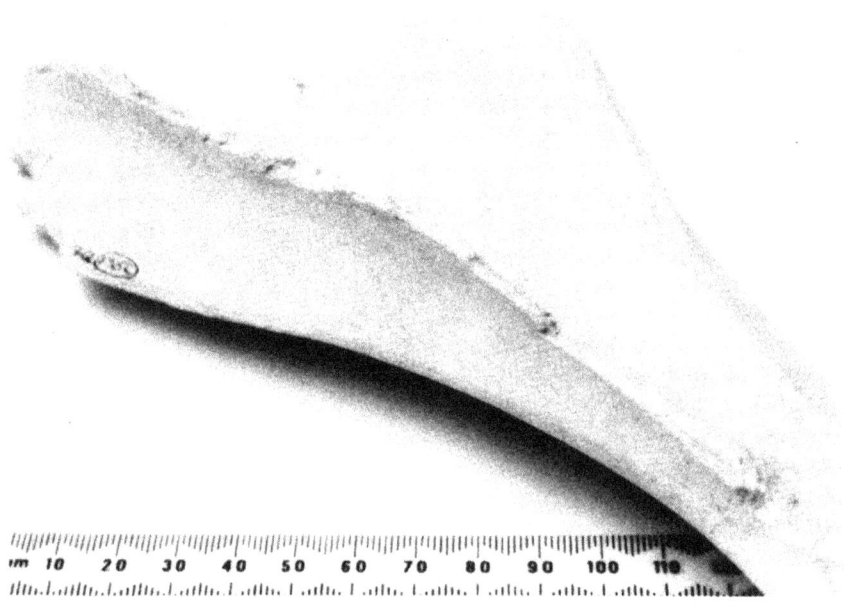

Plate 2 - Cattle scapula from Owslebury showing chop marks running along the lateral spine (Type S4).

meat. Other superficial chop marks on the blade of the scapula may also have been associated with the same filletting process (Types S5-S7).

This technique was one of the few that was commonly encountered on both rural and urban sites (Table 3). However, whereas such marks were almost the only ones associated with the removal of shoulder meat at Winchester, longitudinal and other knife cuts associated with filletting were commonly observed on the blades of scapulae at Owslebury and the other rural settlements (Types S9-S13). Such marks have also been encountered in Iron Age samples in Hampshire. They were found occasionally even in the later Romano-British deposits at Owslebury. Although the poorer preservation of the late Roman assemblage at Owslebury may have resulted in the destruction of a higher proportion of knife cuts, it seems possible that the method was gradually replaced by the use of heavier implements during the Romano-British period (Maltby n.d.3). The use of different types of tools on these rural sites seemingly to carry out the same butchery procedure is interesting and would repay further experimental study. It is possible that the methods varied depending upon whether the meat was removed prior to or after cooking. Alternatively, shoulder meat destined for immediate consumption may have been butchered differently to meat that was to be preserved by salting or smoking.

Humerus

Butchery marks on cattle humeri again varied significantly between urban and rural sites. At Owslebury, marks associated with the disarticulation of the proximal and distal ends from the scapula and radius/ulna respectively almost entirely consisted of knife cuts (Types H14, H10). Knife cuts on shaft fragments were probably associated with filletting (Type H13). Such cuts have commonly been found in earlier Iron Age assemblages. Although knife cuts were found on the medial aspects of seven distal ends of humeri from Winchester Northern Suburns, they formed only a small proportion of the butchery marks (Table 4).

Conversely, several types of butchery marks commonly observed on humeri from Winchester have rarely been encountered on rural sites. This includes the axial chopping through of the bones, usually in an anterio-posterior direction, most commonly observed at the distal articulation (Type H1). In some specimens only the most lateral part of the trochlea and lateral epicondyle had been removed. This may have been the result of blows made during the disarticulation of the elbow joint, using the techniques replicated by Ashdown & Evans (1981: 215). In a few cases, most of the remainder of the humeri remained intact and had not been broken open for marrow. However, the position and severity of the blow was most often such that it also resulted in the splitting open of the shaft, suggesting that marrow extraction was also intended. Indeed, many specimens had been split through the distal articulation more than once (Type H4) and/or broken open from the medial or lateral aspects (Type H9). Shaft fragments also showed evidence of such axial splitting (Types H5-H6). This often resulted in heavy fragmentation of the bones, as illustrated in Plate 3. In some cases such chopping appears to have been too elaborate simply for the extraction of marrow. The bones may have been split into these small pieces during preparation for broth, as van Mensch (1974) has suggested. In a few cases the splintering of the shaft of the bones may have been the result of boneworking, although the humerus is not an ideal shape for such purposes.

Plate 3 - Distal articular surface (trochlea) of two cattle humeri from Winchester Northern Suburbs axially split into small fragments (Type H4).

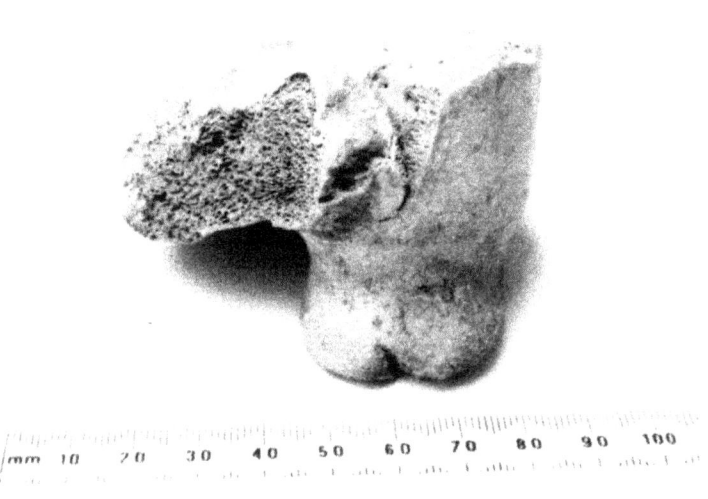

Plate 4 - Cattle humerus from Winchester Northern Suburbs showing typical butchery on the distal medial epicondyle made during disarticulation from the radius/ulna (Type H2).

Another distinctive type of butchery mark encountered in Winchester and Silchester consisted of the slicing off of the most distal part of the medial epicondyle (Type H2 - Plate 4). This would have separated the radius and ulna from the humerus by chopping or sawing through the extensor muscles. The method indicates that the radius and humerus were carefully separated, so that further processing, such as axial splitting, could be carried out on the disarticulated bones. Elsewhere, only two possible parallels were encountered at Owslebury and none at all were observed in any of the other rural samples.

Another common type of butchery mark was confined almost entirely to samples from Winchester and the Manor Farm site in Silchester. This consisted of the removal of small scoops or slices of bone from the shaft (Type H11). Such marks were located mainly on the deltoid ridge and the insertion of the teres major muscle. The marks are consistent with the running of a heavy blade (or possibly a cleaver) axially along the bone to remove meat and other soft tissue. The scoops occurred mainly in areas where the blade encountered natural irregularities on the surface of the bone. This technique, although similar to the one which was commonly used to remove meat from the scapula (Type S4), was not encountered at Owslebury, even in the later Romano-British deposits. One specimen from Brighton Hill South did have several chop marks running along the deltoid ridge, but these were deeper and more irregular than those found in Winchester. The technique seems therefore to have been almost entirely confined to urban sites.

Radius

Variations in butchery patterns were similar to those encountered for the humerus. Thirteen different types of mark were recorded (Table 5). At Winchester, the most common type consisted of splitting the proximal radius in an anterio-posterior direction (Type R1). In some instances this resulted in the removal only of the most lateral part of the articular surface and corresponds with the dismemberment technique which produced similar chops through the lateral part of the trochlea of the humerus. In other cases such chopping seems also to have been designed to open up the marrow cavity. Indeed, often the proximal end was chopped through repeatedly (Types R3-R4). There was also evidence for axial splitting at the distal end of the bone (Type R5). As with the humerus the fragmentation of some specimens was greater than would be necessary simply to gain access to marrow.

Although scoops made by running a heavy blade or occasionally a saw along the surface of the bone to remove meat and other soft tissue were relatively common (Type R14) in the Winchester assemblages, they were not observed as frequently as on the humerus. This is probably because the radius has fewer irregularities along its shaft to cause obstructions to the blade. Similar skims and scoops were found frequently along the posterior aspect of the ulna shaft.

No scoop marks or axially split radii were found at Owslebury, where butchery marks consisted almost entirely of fine knife cuts, even in the later Romano-British layers (Types R11-13).

Pelvic bones

At Winchester and Silchester most butchery on the bones of the pelvis or os coxae consisted of chop marks (Table 6). The most common types were associated with the disarticulation of the femur. This resulted either in the complete severance of the acetabulum (Type P5) or superficial chop or saw marks on or near the joint surface (Type 6). In addition, the shaft of the ilium was often severed in a dorso-ventral/latero-medial direction, (Type P2). This would have detached the caudal part of the pelvis and hindlimb from the vertebral column. Superficial chops on the iliac tuberosity would have been made during the separation of the ilium from the sacrum (Type P1). Various other superficial chop marks were found on ilia from Winchester (Types P3-P4). Chops through the pubis (Type P7) could have been made during the separation of the two halves of the pelvis. Superficial chop or saw marks on the pubis and ischium were also common (Types P8, P10), some of which may have been produced during filletting. So too were the scoop marks made by running a blade along the shafts of the ilium and ischium (Type P11). Such marks were, however, only recorded in the Northern Suburbs sample. Only four specimens from that site bore knife cuts, (Types P12-P16). This was as many as were found in the much smaller sample from the Staple Gardens excavations in Winchester and is an example of the variability of butchery within towns.

At Owslebury, knife cuts on the pelvis outnumbered chop marks in all periods, although the late Roman sample was very small. Butchery marks on the pelvis were rarely encountered in the other rural samples, but knife cuts were the most common. Although most types of chop mark found in the Winchester Northern Suburbs sample were also represented at Owslebury, they were much less frequently observed. Apparently more traditional methods of butchery continued to find favour.

Femur

In many ways variations in butchery marks on the femur mirrored those on the humerus. Axial blade (or occasionally saw) marks associated with filletting left characteristic scoops or skims on various protuberances of the shaft (Type F9). The most common locations for these included the lesser trochanter and the medial and lateral supracondyloid tuberosities; 34% of the femur fragments from Winchester Northern Suburbs bore such marks. They were also frequently encountered at Staple Gardens but were not observed on any of the 18 fragments from the Silchester South Gate excavations. However, two specimens were found in the Manor Farms deposits (Table 7). Only one possible parallel was found at Owslebury in a late Roman context.

Axial splitting of the femur was also common in the Northern Suburbs but was not observed in any of the other samples (Types F3-F6). The severe fragmentation is similar to that on the humerus and radius. The very splintered fragments were usually in layers associated with boneworking.

Many of the proximal articular surfaces at Winchester were chopped through separating them from the rest of the bone (Type F1). This type of butchery corresponds to the chops through the acetabulum (Type P5) made during the severing of the hindlimb from the pelvis. Less frequently, the distal articular surface of the femur had been chopped through horizontally (Type F6) and in a few cases the medial and lateral condyles had been chopped through axially in a medio-lateral plane (Type F14).

Occasionally examples of chop marks were encountered amongst the rural assemblages but the majority of the butchery marks consisted of various knife cuts located on different parts of the bone (Types F10-F13).

Tibia

The vast majority of butchery marks on cattle tibiae from the Winchester sample consisted of either superficial axial blade marks made during filletting or axial chops made during the splitting of the bones (Table 8). Twenty one per cent of the identifiable tibia fragments from the Northern Suburbs excavations bore blade marks (Type T12). Plate 5 illustrates an example of the typical slicing of the proximal anterior border made during this process. Other skims or scoops were commonly located on the medial and lateral borders along the length of the bones. No such marks were found in the Silchester samples, although two similar occurrences were recorded in the later Romano-British assemblage from Owslebury.

Axial chops through the tibiae usually ran in an anterio-posterior direction (Types T2, T4, T6). In some cases, as with the other upper limb boness, the procedure was repeated (Types T3, T5, T7), possibly either to prepare the bones for broth boiling, or in some instances to obtain suitable splinters of the shaft for boneworking. Axially split tibia fragments were also found in both samples from Silchester but not on any of the rural sites.

At Owslebury most of the butchery marks consisted of knife cuts associated with the filletting of meat from the shaft (Type T10). Marks associated with the disarticulation of the tibia from its adjacent bones were rare on all sites.

Astragalus

Once again a clear distinction between the urban and rural samples can be seen (Table 9). Knife cuts on the anterior surface of these bones, usually on the projecting medial and lateral borders, were the principal butchery marks encountered in the rural samples (Types A9-A10). These resulted from the disarticulation of the tibia from the limb extremities and were also common in Iron Age assemblages from Hampshire. They were not found, however, in the Winchester and Silchester samples. There, much heavier chopping was employed to disarticulate the bones. Sometimes this resulted in the bones being completely severed horizontally (Types A1, A3, A5). Blows were usualy delivered from the front. At Owslebury, although a few chop marks were encountered, most were only superficial and essentially no different from most of the knife cuts.

A few of the astragali from Winchester had been split axially (Types A7-A8). This may have happened during the splitting of the tibia, in cases where the bones were still articulated. A few specimens bore axial blade marks similar to those found more frequently on the long bones (Type A11).

Calcaneus

In the Winchester Northern Suburbs sample, chop and saw marks were frequently encountered on the calcaneal tuberosity, particularly on the medial aspect. Such butchery often severed the most proximal part of the bone (Types C1-C2 - Plate 6). Chop marks were also found commonly on other parts of the calcaneus in the urban assemblages (Types C3-C6). These frequently went completely through the bone (Types C3 and C5). Most were probably associated with the disarticulation of the tibia from the extremities, although a few specimens may have been broken for further processing. Only one knife cut was observed on a calcaneus from Winchester, whereas this was the only type of butchery mark found on specimens from Owslebury (Table 10).

Centroquartal

The centroquartal or navicular-cuboid lies between the astragalus and metatarsal. Most butchery marks found on it were associated with the separation of the metatarsal from the upper limb bones. However, the tools used and the location of the butchery marks varied significantly between Winchester Northern Suburbs and Owslebury (Table 11). In the former sample, superficial chop marks were found mainly on the medial and lateral borders of the posterior aspect of the bone (Type Q2). At Owslebury, none of the specimens bore evidence of chop marks and nearly all the knife cuts were located on or near the anterior surface (Type Q4). Three bones from Winchester Northern Suburbs had been chopped through axially, probably when still attached to the metatarsus when that was split (Type Q1).

Metapodials

Butchery marks on cattle metapodials were less common than on the upper limb bones. In Winchester this was largely because axial chop marks and blade marks made during filletting, although present, were not found in great numbers (Types M1-M4, M14 - Tables 12-13). Axial splitting tended to be restricted to certain features and may have been mainly associated with boneworking activities. The numbers of distal ends that had been chopped through obliquely (Type M3) are minimum figures, since 14 more distal articulations of metapodials butchered in this manner were too fragmentary to determine whether they belonged to metatarsals or metacarpals.

None of the metapodials from Silchester were split longitudinally, although they were consistently broken horizontally, presumably to extract marrow (Maltby 1984b: 202). The practice was also rarely encountered on the rural sites, although a few metacarpals did bear axial chop marks through the proximal end (Type M1 - Table 13). As usual, the rural sites produced most of the examples of knife cuts on these bones (Types M10-M13). Butchery evidence associated with the disarticulation of the metatarsal from the upper hind limb usually consisted of superficial chop marks located on the medial and lateral borders of the posterior surface at Winchester (Type M5), whereas knife cuts mainly on the anterior surface of specimens were most common on the rural sites. These correspond to the patterns observed on the centroquartal bones.

Axial blade marks associated with filletting were found on only four of the metatarsal fragments from Winchester Northern Suburbs (Type M14). The

Plate 5 – Cattle tibia from Winchester Northern Suburbs showing typical blade marks on the proximal anterior border made during defleshing (Type T12).

Plate 6 – Cattle calcaneus from Winchester Northern Suburbs showing chop and saw marks on the lateral aspect of the calcaneal tuber. The most proximal part of the bone has been completely cut off (Type C1-C2).

practice appears to have been rarely carried out on these bones, which have very little meat adhering to them.

First phalanx

These bones produced more knife cuts than chop marks even in the urban samples (Table 14). This was because it was common practice to make incisions around the toes to enable the skins to be removed. Such marks were found commonly on all but the axial aspects of the shafts of the bones (Types Ph4-Ph6). However, there were variations between the samples. Chop marks were confined almost exclusively to the samples from Winchester (Types Ph8-Ph10) and there was a higher proportion of butchery marks near the proximal articulation associated with the disarticulation of the phalanges from the distal metapodials in the urban samples (Types Ph1-Ph3, Ph7).

Skull

Butchery marks on skull fragments will not be discussed in detail here, since a wider range of samples is required. Generally, rural sites again produced more evidence for the use of knives. At Owslebury, only 22 skull fragments bore chop marks, whereas 89 possessed knife cuts. This pattern was consistent throughout the Romano-British deposits at that site. Most of the chop and saw marks were associated with the removal of the horn and horn core from the skull. In the Winchester Northern Suburbs sample only one fragment of maxilla bore a knife cut. The remaining butchery marks consisted of chop marks, which were recorded on 79 of the skull fragments, most of which were associated with the removal of the hore core from the frontal. However, a cleaver was also commonly used to disarticulate the skull from the mandibles, as chop marks on the temporal and zygomatic show, and from the neck, as demonstrated by chops on and around the occipital condyles. It should be noted, however, that one of the two large horn core concentrations was found on the rural site of Brighton Hill South; the other was in the pit in Winchester Northern Suburbs described above. Distinctive sawing marks on many of the horn core fragments were made during the removal of the horn sheath (Maltby n.d.6). This demonstrates that specialised activities associated with the processing of cattle carcases were not totally confined to the towns.

Vertebrae and ribs

Analysis of butchery on these bones is restricted because the ribs in particular are not always identifiable to species. Once again, however, the urban samples produced more evidence for the use of heavier implements than the rural samples. Lumbar and thoracic vertebrae were commonly chopped through to remove the lateral processes and the flanks of the animal. A similar procedure was commonly encountered on the rural settlements but the butchery tended to remove less of the vertebral bodies. Midline splitting of the vertebrae was comparatively rare but a few specimens from Winchester had such butchery. Some thoracic and lumbar vertebrae from those deposits had also been chopped through transversely during segmentation of the spinal column. Such specimens were rarely encountered on the rural sites. Similarly, ribs were commonly segmented into relatively short lengths in Winchester.

DISCUSSION

It is clear therefore that the implements and methods of butchering used on cattle carcases on these Hampshire sites varied considerably. The urban samples from Winchester and Silchester generally produced bones that had been butchered mainly with cleavers or similar chopping implements, whereas knives were still commonly employed on the rural settlements, even in the later Romano-British period. The impression is that the carcases were more heavily and consistently butchered in the two towns than in the rural settlements. In particular, more of the limb bones were broken into small fragments, in order to exploit marrow and grease from the bones. To quantify this is difficult because fragmentation can also be caused by scavenging animals, trampling and weathering. Table 15 shows the fragment sizes of the major limb bones identifiable to cattle at Owslebury and Winchester. The average size of the fragments in Winchester tended to be smaller despite the fact that the assemblages from Owslebury had suffered greater fragmentation from erosion and canid scavenging.

Cattle were by far the most important producers of meat. It is therefore to be expected that butchers in Roman towns would intensively exploit their carcases in order to maximise their food value. Consequently highly intensive techniques of carcase processing new to Britain are more likely to be encountered in the towns. On rural settlements butchery was unlikely to have been a full-time specialist activity. Traditional butchery practices were still in existence even in the later Roman period at Owslebury. Admittedly, some specimens from there and the other rural settlements bore butchery marks similar to those found in large numbers in Winchester, suggesting that some of the new methods of carcase processing were copied in the country. Such marks were, however, not found consistently in the rural samples, and certain types of chop mark were not observed at all. Although, as Done (1986: 145) has pointed out, the butchering methods practised in the towns would appear to lack finesse in the use of choppers instead of knives, it is plausible that such butchery methods were better suited to butchering a lot of carcases quickly.

This survey has been limited to Hampshire sites, and a general survey of Romano-British butchery practices is premature given the limited nature of most discussions of Romano-British butchery in published excavation reports. There are, however, a few sites where particular styles of butchery have been reported or appear to be absent.

Axial chops through limb bones

Chops through the lateral aspects of the distal articulation of the humerus and the proximal articulations of the radius and ulna have been recorded on several sites. In some cases in Winchester, these chops appear to have been made during the disarticulation of the elbow joint. Ashdown & Evans (1981) replicated such butchery marks which they had encountered in Late Iron Age samples from the Skeleton Green excavations at Puckeridge, Hertfordshire. Since this settlement appears to have been partly inhabited by literate Romans or Romanized Gauls engaged in intensive trading activities (Partridge 1981: 351), it is tempting to speculate that these included specialist butchers who imported their butchery techniques. However, we need to compare earlier Iron Age and other contemporary samples from South-East England to test this theory. Certainly similar butchery evidence has not been found in the Middle or Late Iron Age samples from Hampshire but has been noted in early Roman deposits at Winchester and Silchester. Elsewhere in Hampshire, axial chops through the lateral condyles of the humeri were observed in assemblages from the small town at Neatham (Done 1986: 144).

In Cirencester, evidence for the axial splitting of the elbow joint has also been encountered in early Roman deposits (Thawley 1982: 219-20). These were military deposits and there is the possibility that many of these butchering techniques may have been imported and spread by butchers connected with provisioning the armed forces with their meat supplies. Once established, the techniques continued to be used in the towns and possibly on other "Romanized" settlements.

In the Northern Suburbs and Staple Gardens samples from Winchester, however, axial chopping of the limb bones was also carried out to split open the marrow cavity of the bones. This was a method that was not found as frequently in the samples from the Silchester Defences. It was also rare at Fishbourne (Grant 1971). We have already noted that only very rare examples of split limb bones were encountered on rural sites in Hampshire. At the Saxon shore fort at Brancaster, Norfolk, it was noted that although some radii had been split axially, distal tibiae were usually found intact (Jones et al. 1985: 146). Some examples of split upper limb bones were recorded in the detailed analysis of butchery marks made on a sample of bones from excavations in Lincoln (Aird 1985: 13-14). However, it appears that evidence for this practice will not be encountered in every urban assemblage. The only reported samples in Britain in which axially split upper limb bones appear to have been found as commonly as in the Winchester samples are the ones from the Tower Street excavations in Cirencester (Maltby 1984a: 130-132) and the Balkerne Street excavations in Colchester, Essex (Luff 1982: 104). Repeated axial chopping and splitting was also common on several Roman military and urban sites in Europe. It has been argued that such butchery was carried out during large-scale specialised processing of the bones for broth production (van Mensch 1974). It may have been a procedure that was carried out only by certain butchers in the larger towns or military installations.

Superficial axial blade marks

These marks associated with filletting were very common on the major meat-bearing limb bones in the samples from Winchester. However, if we exclude the scapula from the analysis, there is less consistency about their frequency in some of the other urban samples. In the assemblages from Silchester, they were encountered only in some of the Manor Farm deposits (Maltby 1984b). In Cirencester, such marks were noted on bones from the early military deposits (Thawley 1982) and in the later Tower Street layers (Maltby 1984a: 132). They were also observed in an assemblage from Lincoln (Aird 1985: 14-15). However, even in the reports which give details of butchery, observations of this distinctive style of processing do not appear to have been made in any of the other urban or rural samples. This may partially be a consequence of a lack of clarity or detail in the reports but it also raises the possibility that this type of butchery was again restricted to certain centres and was only carried out by particular butchers or in certain circumstances.

By contrast with the apparently restricted distribution of such butchery marks on the humerus, radius, femur and tibia, filletting marks which removed part or all of the spine or the scapula have been reported on a much wider range of sites. In addition to the urban and rural samples from Hampshire, already discussed, such butchery has been noted at Cirencester (Thawley 1982: 219), Portchester Castle (Grant 1975: 392), Brancaster (Jones et al. 1985: 144), the fort of Longthorpe (Marples 1974: 123) and Colchester (Luff 1982: 100). Contrasts between the sites regarding this style of butchery may be found in the frequency that such methods were used rather than simply noting their presence or absence. It has been shown that in several of the rural samples other types of filletting marks were also encountered, and butchery involving the removal of the spine was found less consistently than in the urban samples (see also Maltby 1985b: 24-5).

CONCLUSIONS

This paper has shown that detailed recording and analysis of butchery marks can provide a great deal of information relating to how animal carcases were processed by urban butchers. The analysis of the Hampshire sites has demonstrated that such work can reveal a great deal of variability. It has shown that butchers in the major Roman towns of Silchester and Winchester processed cattle carcases in ways different to the methods traditionally employed in the area. Since major towns such as these probably needed to import large numbers of cattle to provide meat for the inhabitants, the role of the specialist butcher was an important one. Consequently, new, probably faster, methods of processing were introduced. At the same time, the exploitation of the carcases became more intensive. Centralisation and specialisation meant that by-products such as marrow, grease and the raw materal for boneworking could be produced in bulk. On small rural settlements such procedures may not have been cost-effective because of the relatively small amount of carcase processing that would have taken place. It remains to be seen how widely such butchery methods spread to sites such as villas and small towns. It also seems that even in the large towns, not all the processes were carried out on every carcase. Future analysis of urban samples may be able to distinguish the spread, ubiquity, and distribution of particular processing techniques.

It has been intimated in the above discussion that variations in butchery practices are likely to result from a combination of cultural, technological and logistical considerations. The Romano-British period with its complex political, economic and social networks appears to be one where such variations were particularly marked. By establishing the scale, scope and distribution of different methods of carcase processing in various types of settlement throughout Roman Britain, butchery analysis may be able to assist in the understanding of how these networks operated.

Acknowledgements

I would like to thank the excavators who gave their permission to include data from unpublished excavations in this paper. English Heritage funded the original analysis of the faunal data. Steve Shrimpton of the Southampton University Teaching Media Unit took the photographs for the plates.

Sample	K J1	K 2	K 3	K 4	C 5	C 6	C 7	K 8	K 9	K 10	C 11	C 12	C 13	K 14	C 15	F
Winchester, N.S.	1				5	5	7	2		2	12	38	7		2	417
Staple Gardens		2			3						3	2			1	46
Silchester, S.G		2						1			4	1				65
Manor Farm							(none)									9
Owslebury 1 A.D.	37	2	1		1			35	14	6	14	4	3	8		767
1-2 A.D.	5	1	2		1			8	2	4	4	1		3	1	366
3-4 A.D	11	2						2			3			1		631
Winnall Down	2							1	1		2			1		83
Cowdery's Down	1		1			1	1			3	1		2	1		63
Abbotstone Down	3							3								43
Brighton Hill S.	3							2	1		2					131
Little Somborne	1															42
Balksbury, 1973	1							2			2					59

K = knife cut(s)
C = predominantly chop mark(s)
F = total number of mandible fragments
J1 = dorso-ventral (or oblique) knife cuts - lateral diastema
J2 = dorso-ventral (or oblique) knife cuts - medial diastema
J3 = cranio-caudal knife cuts - lateral diastema
J4 = cranio-caudal knife cuts - medial diastema
J5 = dorso-ventral (or oblique) chop/saw marks - lateral diastema
J6 = dorso-ventral (or oblique) chop/saw marks - medial diastema
J7 = dorso-ventral/cranial-caudal chop through medial diastema
J8 = cranio-caudal knife cuts - lateral ramus near condyle
J9 = other knife cuts - caudal part of ramus near condyle
J10 = knife cuts - other parts of ramus
J11 = cranio-caudal chop marks - lateral ramus near condyle
J12 = chop/saw marks - caudal ramus on and below condyle
J13 = chop/saw marks - other parts of ramus
J14 = knife cuts below cheek tooth row (mostly on lateral)
J15 = superficial chop marks below cheek tooth row

Table 2. Butchery marks on cattle mandibles

| | C | C | C | B | B | B | C | K | K | K | K | K | K | |
Sample	S1	2	3	4	5	6	7	8	9	10	11	12	13	F
Winchester, N.S.	40	4	2	54	3	4	5		2		1			324
Staple Gardens	6		1	6	1	1	2							48
Silchester, S.G	1			4	1									22
Manor Farm			1	1										5
Owslebury 1st A.D.			1	15	2			11	11		7	9	3	215
1-2 A.D.	1		1	7		2		1	7	2	1	1		79
3-4 A.D.	1			14	4	3		1	2	1	1	2		192
Winnall Down				2	1			3	1			1		42
Cowdery's Down			1	3								1		13
Abbotstone Down				3								1		20
Brighton Hill S.				2				2	1	1				40
Little Somborne				(none)								19
Balksbury, 1973							1				1			17

K = knife cut(s)
C = predominantly chop mark(s)
B = predominantly heavy blade mark(s)
F = total number of scapula fragments
S1 = axial/oblique chop through glenoid cavity running in latero-medial direction
S2 = repeated axial/oblique chops through glenoid cavity running in latero-medial direction
S3 = horizontal superficial chop marks - around glenoid cavity
S4 = axial chop/blade/saw marks - lateral spine
S5 = other axial chop/blade/saw marks - lateral aspect of blade
S6 = superficial chop/blade marks - medial/caudal aspects of blade
S7 = other chop/blade/saw marks - lateral aspect of blade
S8 = horizontal knife cuts - around glenoid cavity
S9 = axial knife cuts - lateral aspect of blade (including spine)
S10 = axial knife cuts - medial blade
S11 = other knife cuts - lateral and cranial aspects of blade
S12 = other knife cuts - medial and caudal aspects of blade
S13 = knife cuts - near proximal end

Table 3.   Butchery marks on cattle scapulae.

Sample	C H1	C 2	C 3	C 4	C 5	C 6	C 7	C 8	C 9	K 10	B 11	C 12	K 13	K 14	F
Winchester, N.S.	52	39	12	21	20	2	3		4	7	63	4			324
Staple Gardens	7	5			2		2	1			10	3	1		48
Silchester, S.G	2	3	2	1		1				1		1			22
Manor Farm				1	1						1				5
Owslebury 1st A.D.		1				1			1	19		2	16	5	215
1-2 A.D.										7			2		79
3-4 A.D.		1					1			8		1	2	2	192
Winnall Down										5					42
Cowdery's Down										1					13
Abbotstone Down										1		1	1		20
Brighton Hill S.										5	1		3		40
Little Somborne							1		1						19
Balksbury, 1973								2		2					17

K = knife cut(s)
C = predominantly chop marks
B = predominantly heavy blade or chop marks
F = total number of humerus fragments
H1 = axial chop through distal articulation (trochlea) running in anterio-posterior direction
H2 = horizontal/oblique chop through distal surface of medial epicondyle
H3 = axial/oblique chop through proximal articulation
H4 = repeated axial chops through distal articulation running in anterio-posterior direction
H5 = axial/oblique chop through shaft running in anterio-posterior direction
H6 = repeated axial/oblique chops through shaft
H7 = oblique/anterio-posterior superficial chop/saw marks - medial aspect distal articulation
H8 = superficial chop/saw marks - proximal end
H9 = axial/oblique chop through medial or lateral aspects of distal articulation
H10 = knife cuts - distal end (mostly on medial aspect)
H11 = superficial axial blade/chop/saw marks on shaft
H12 = other superficial chop/saw marks on shaft
H13 = knife cuts on shaft
H14 = knife cuts - proximal end

Table 4.   Butchery marks on cattle humeri.

Sample	C R1	C 2	C 3	C 4	C 5	C 6	C 7	C 8	C 9	C 10	K 11	K 12	K 13	B 14	F
Winchester, N.S.	73	4	23	8	22		9	1	5	4				19	316
Staple Gardens	6		5	1		1				1				11	51
Silchester, S.G	4														10
Manor Farm	1														4
Owslebury 1st A.D.											16	1	9		156
1-2 A.D.							1		1		4	2	1		74
3-4 A.D.											6		2		184
Winnall Down							1				1	1			34
Cowdery's Down											1				8
Abbotstone Down							(none)								16
Brighton Hill S.									1						21
Little Somborne							(none)								20
Balksbury, 1973					1						4				13

K = knife cut(s)
C = predominantly chop mark(s)
B = predominantly heavy blade mark(s)
F = total number of radius fragments
R1 = axial chop through proximal articulation running in anterio-posterior direction
R2 = axial chop through proximal articulation running in medio-lateral direction
R3 = repeated axial chops through proximal articulation running in anterio-posterior direction
R4 = axial chops through proximal articulation running in medio-lateral and anterio-posterior directions
R5 = axial chop through distal articulation running in anterio-posterior direction
R6 = superficial chop/saw marks on shaft
R7 = axial chop through shaft running in anterio-posterior direction
R8 = repeated axial chops through shaft running in anterio-posterior direction
R9 = superficial horizontal chop marks - proximal medial aspect
R10 = horizontal chop marks (mostly superficial) - distal end
R11 = knife cuts - proximal end (mostly on medial aspect)
R12 = knife cuts - distal end
R13 = knife cuts on shaft
R14 = superficial axial blade/chop/saw marks on shaft

Table 5. Butchery marks on cattle radii.

Sample	C P1	C 2	C 3	C 4	C 5	C 6	C 7	C 8	C 9	C 10	B 11	K 12	K 13	K 14	K 15	K 16	F
Winchester, N.S.	10	13	8	8	24	14	5	7		13	11	1		1	1	1	229
Staple Gardens	2	4	1	1	5	2		1	2			1		2	2		34
Silchester, S.G			1		8	1	1	2		1					1		40
Manor Farm							(none)										4
Owslebury 1 A.D.	1	1		2	2	1		2				5	2	4	3	1	169
1-2 A.D.		3			1			1					2			6	71
3-4 A.D.				2								2			2		166
Winnall Down					1					1			1	1		1	32
Cowdery's Down							(none)										7
Abbotstone Down														1			12
Brighton Hill S.															2		25
Little Somborne							(none)										19
Balksbury, 1973							(none)										23

K = knife cut(s)
C = predominantly chop mark(s)
B = predominantly heavy blade mark(s)
F = total number of pelvic fragments
P1 = chop/saw marks - iliac tuberosity (articular surface with sacrum)
P2 = dorso-ventral/latero-medial chop through shaft of ilium
P3 = superficial dorso-ventral chop/saw marks - shaft of ilium
P4 = other superficial chop/saw marks - shaft of ilium
P5 = chop through acetabulum
P6 = superficial chop/saw marks in and around acetabulum
P7 = cranio-caudal/oblique chop through shaft of pubis
P8 = superficial chop/saw marks - shaft of pubis
P9 = chop through shaft of ischium
P10 = superficial chop/saw marks - shaft of ischium
P11 = superficial blade/chop/saw marks - shafts of ilium and ischium
P12 = knife cuts - lateral aspect of shaft of ilium
P13 = other knife cuts - ilium
P14 = knife cuts - in and around acetabulum
P15 = knife cuts - pubis
P16 = knife cuts - ischium

Table 6.   Butchery marks on cattle pelvic bones.

Sample	C F1	C 2	C 3	C 4	C 5	C 6	C 7	C 8	B 9	K 10	K 11	K 12	K 13	C 14	F
Winchester, N.S.	21	1	5	13	22	5		6	100					4	291
Staple Gardens	2								16						49
Silchester, S.G	1					1									22
Manor Farm									2						7
Owslebury 1st A.D.	1									3		7	4		168
1-2 A.D.										2	1	1	1		77
3-4 A.D.			1			1		1				3			173
Winnall Down									1			2	1		36
Cowdery's Down	1														18
Abbotstone Down	1														9
Brighton Hill S.						(none)									23
Little Somborne						(none)									13
Balksbury, 1973											1	1			12

K = knife cut(s)
C = predominantly chop mark(s)
B = predominantly heavy blade mark(s)
F = total number of femur fragments
F1 = proximal articular surface (head and neck) chopped through
F2 = superficial chop marks on and around proximal head and neck
F3 = axial chop through proximal running in anterio-posterior direction
F4 = axial/oblique chop through shaft running in anterio-posterior direction
F5 = axial/oblique chop through distal running in anterio-posterior direction
F6 = repeated axial/oblique chops through distal running in anterio-posterior direction
F7 = superficial horizontal chop/saw marks - distal end
F8 = horizontal chop through distal articular surface
F9 = superficial axial blade/chop/saw marks on shaft
F10 = knife cuts - medial aspect of proximal end
F11 = other knife cuts - proximal end
F12 = knife cuts on shaft
F13 = knife cuts - distal end
F14 = axial chop through distal lateral and medial condyles running in medio-lateral direction

Table 7. Butchery marks on cattle femora.

Sample	C T1	C 2	C 3	C 4	C 5	C 6	C 7	C 8	K 9	K 10	K 11	B 12	C 13	F
Winchester,N.S.	1	19	3	22	5	34	10			1		81	3	381
Staple Gardens	1	1		1		3	1			1		13	3	61
Silchester, S.G						1								24
Manor Farm						1	1							14
Owslebury 1st A.D.									1	7	2		1	179
1-2 A.D.										4				70
3-4 A.D.										5		2	1	216
Winnall Down						(none)								31
Cowdery's Down						(none)								8
Abbotstone Down						(none)								18
Brighton Hill S.										1				40
Little Somborne										1				19
Balksbury, 1973								1						17

K = knife cut(s)
C = predominantly chop mark(s)
B = predominantly heavy blade mark(s)
F = total number of tibia fragments
T1 = superficial horizontal/oblique chop/saw marks - proximal end
T2 = axial chop through proximal usually running posterio-anteriorly
T3 = repeated axial chops through proximal
T4 = axial chop through shaft
T5 = repeated axial chop through shaft
T6 = axial chop through distal usually running posterio-anteriorly
T7 = repeated axial chop through distal
T8 = superficial horizontal chop/saw marks - distal end
T9 = knife cuts - proximal end
T10 = knife cuts on shaft
T11 = knife cuts - distal end
T12 = superficial axial blade/chop/saw marks on shaft
T13 = other superficial chop/saw marks on shaft

Table 8.   Butchery marks on cattle tibiae.

Sample	C A1	C 2	C 3	C 4	C 5	C 6	C 7	C 8	K 9	K 10	B 11	F
Winchester, N.S.	4	14	10	6	5	2	4	1			3	74
Staple Gardens	2	2	2	3	1		1				2	12
Silchester, S.G	3		1	3	1							9
Manor Farm					1							2
Owslebury 1st A.D.		1							12	5		57
1-2 A.D.									6	2		22
3-4 A.D.				3	1				1	3		91
Winnall Down			1						3			22
Cowdery's Down					(none)							3
Abbotstone Down									1			5
Brighton Hill S.					(none)							7
Little Somborne									1	2		5
Balksbury, 1973		1							3	1		11

K = knife cut(s)
C = predominantly chop mark(s)
B = predominantly heavy blade mark(s)
F = total number of astragalus fragments
A1 = oblique/horizontal chop through proximal usually running anterio-posteriorly
A2 = superficial chop/saw marks - proximal
A3 = oblique/horizontal chop through centre of bone usually running anterio-posteriorly
A4 = superficial anterio-posterior chop/saw marks - anterior aspect of central part of bone
A5 = oblique/horizontal chop through distal usually running anterio-posteriorly
A6 = superficial chop/saw marks - distal
A7 = axial/oblique split through bone
A8 = repeated axial/oblique splits through bone
A9 = knife cuts - anterior aspect of centre of bone
A10 = knife cuts - anterior aspect distal end
A11 = superficial axial blade/chop/saw marks

Table 9. Butchery marks on cattle astragali.

Sample	C1	C2	C3	C4	C5	C6	K7	K8	F
Winchester, N.S.	23	4	2	1	11	3	1		87
Staple Gardens		2			1				13
Silchester, S.G	1	1			1				11
Manor Farm					1				4
Owslebury 1st A.D.							1	3	56
1-2 A.D.							1	3	27
3-4 A.D.								3	99
Winnall Down								2	16
Cowdery's Down						1			4
Abbotstone Down				(none)					9
Brighton Hill S.				(none)					13
Little Somborne					2				3
Balksbury, 1973				(none)					9

K = knife cut(s)
C = predominantly chop mark(s)
F = total number of calcaneus fragments
C1 = oblique/medio-lateral chops through calcaneal tuber
C2 = superficial chop/saw marks - calcaneal tuber
C3 = oblique/horizontal chops through distal end
C4 = superficial chop/saw marks - distal end
C5 = oblique/horizontal chops through centre of bone
C6 = superficial chop/saw marks - centre of bone
C7 = knife cuts - centre of bone
C8 = knife cuts - distal end

Table 10. Butchery marks on cattle calcanea.

Sample	Q1	C2	C3	K4	K5	F
Winchester, N.S.	3	9		1		27
Staple Gardens		(none)				1
Silchester, S.G		(none)				-
Manor Farm		(none)				-
Owslebury 1st A.D.				7		26
1-2 A.D.				5		22
3-4 A.D.				6	1	51
Winnall Down				2		11
Cowdery's Down		(none)				-
Abbotstone Down		(none)				-
Brighton Hill S.		(none)				2
Little Somborne		(none)				2
Balksbury, 1973		(none)				-

K = knife cut(s)
C = predominantly chop mark(s)
Q1 = axial chop through bone
Q2 = superficial chop/saw marks - posterior/lateral surfaces
Q3 = superficial chop/saw marks - anterior surface
Q4 = knife cuts - anterior suface (+ lateral and medial)
Q5 = knife cuts - posterior surface

Table 11. Butchery marks on cattle centroquartals.

Sample	C M1	C 2	C 3	C 4	C 5	C 6	C 7	C 8	C 9	K 10	K 11	K 12	K 13	B 14	F
Winchester, N.S.	9	2	11	8	13		2	2	1		1	1	1	4	260
Staple Gardens	1	1	1	1	1				2						30
Silchester, S.G											2				50
Manor Farm							(none)								3
Owslebury 1st A.D.										1	6	1	1	4	209
1-2 A.D.											3		1		103
3-4 A.D.							1			1	2		1		197
Winnall Down										1	3				49
Cowdery's Down							(none)								7
Abbotstone Down											1		2		18
Brighton Hill S.							(none)								33
Little Somborne				1							1				13
Balksbury, 1973							(none)								22

K = knife cut(s)
C = predominantly chop mark(s)
B = predominantly heavy blade mark(s)
F = total number of metatarsal fragments
M1 = axial chop through proximal end
M2 = axial chop through shaft
M3 = axial/oblique chop through distal end
M4 = repeated axial chops through proximal and shaft
M5 = superficial medio-lateral chop/saw marks - posterior aspect of proximal end
M6 = superficial medio-lateral chop/saw marks - anterior aspect of proximal end
M7 = superficial horizontal chop/saw marks - shaft
M8 = horizontal chop/saw through shaft
M9 = horizontal chop/saw marks - distal end
M10 = medio-lateral knife cuts - anterior aspect of proximal end
M11 = medio-lateral knife cuts - posterior aspect of proximal end
M12 = knife cuts on shaft
M13 = knife cuts - distal end
M14 = superficial axial blade/chop/saw marks on shaft

Table 12.   Butchery marks on cattle metatarsals.

Sample	CM1	C2	C3	C4	C5	C6	C7	C8	C9	K10	K11	K12	K13	B14	F
Winchester, N.S.	6	2	4	2		1		1					1		143
Staple Gardens	4	4									1		2		39
Silchester, S.G						(none)									41
Manor Farm						(none)									4
Owslebury 1st A.D.							1			3	1	1	4	1	164
1-2 A.D.	1									2		2		2	78
3-4 A.D.													2		138
Winnall Down	2														30
Cowdery's Down						(none)									11
Abbotstone Down							1								13
Brighton Hill S.						(none)									17
Little Somborne	1														12
Balksbury, 1973						(none)									20

F = total number of metacarpi fragments
Otherwise Key = Table 12

Table 13.   Butchery marks on cattle metacarpals.

Sample	KPh1	K2	K3	K4	K5	K6	K7	C8	C9	C10	F
Winchester, N.S.		10	1	10	7	15	1	7	2	5	181
Staple Gardens		3					1				15
Silchester, S.G	1	1	1		8		1				39
Manor Farm					(none)						5
Owslebury 1st A.D.	2	3	1	7	2	10	2	1			107
1-2 A.D.				3	1	8					69
3-4 A.D.	1		1	7	1	5	1				143
Winnall Down				2		2					18
Cowdery's Down				1	1	1					3
Abbotstone Down				1		1	1				7
Brighton Hill S.					(none)						2
Little Somborne					1						7
Balksbury, 1973					(none)						1

K = knife cut(s)
C = predominantly chop mark(s)
F = total number of fragments
Ph1 = medio-lateral knife cuts - anterior aspect of proximal
Ph2 = medio-lateral knife cuts - posterior aspect of proximal
Ph3 = anterio-posterior knife cuts - peripheral aspect of proximal
Ph4 = medio-lateral knife cuts - anterior aspect of shaft
Ph5 = medio-lateral knife cuts - posterior aspect of shaft
Ph6 = anterio-posterior knife cuts - peripheral aspect of shaft
Ph7 = knife cuts - distal
Ph8 = superficial medio-lateral chop/saw marks - posterior aspect of proximal
Ph9 = superficial medio-lateral chop/saw marks - anterior aspect of proximal
Ph10 = superficial chop/saw marks - posterior aspect of shaft

Table 14   Butchery marks on cattle first phalanges.

		Percentage of Complete Bone					Mean
Bone	Sample	100%	75%	50%	25%	<25%	Size*
Humerus	WNS		14	7	55	271	.16
	OWS 1 A.D.	9	23	37	60	126	.28
	OWS 1-2 A.D.	3	5	8	23	40	.26
	OWS 3-4 A.D.	7	31	21	36	112	.29
Radius	WNS	8	5	6	46	251	.16
	OWS 1 A.D.	18	15	10	65	48	.35
	OWS 1-2 A.D.	8	4	5	23	34	.31
	OWS 3-4 A.D.	22	22	16	42	87	.35
Femur	WNS		5	2	39	253	.13
	OWS 1 A.D.	10	5	5	48	106	.22
	OWS 1-2 A.D.	6	7	2	20	42	.28
	OWS 3-4 A.D.	15	23	11	25	100	.31
Tibia	WNS	2	7	4	74	292	.15
	OWS 1 A.D.	12	14	15	56	72	.30
	OWS 1-2 A.D.	3	6	5	28	28	.28
	OWS 3-4 A.D.	14	49	11	43	104	.35
Metacarpal	WNS	15	8	9	65	46	.32
	OWS 1 A.D.	18	20	11	58	56	.36
	OWS 1-2 A.D.	7	8	2	31	28	.32
	OWS 3-4 A.D.	15	28	7	40	48	.39
Metatarsal	WNS	17	11	28	71	133	.27
	OWS 1 A.D.	9	26	16	63	95	.30
	OWS 1-2 A.D.	5	5	9	32	51	.26
	OWS 3-4 A.D.	28	25	10	35	99	.36

WNS = Winchester Northern Suburbs (3rd-4th Centuries A.D.)
OWS = Owslebury
* Mean size calculated by dividing proportion of each bone represented by total number of fragments. Fragments of <25% were given a value of .10

Table 15. Fragmentation data for cattle limb bones from Winchester Northern Suburbs and Owslebury.

REFERENCES

AIRD P.M. (1985) On distinguishing butchery from other post-mortem destruction: a methodological experiment applied to a faunal sample from Roman Lincoln. In <u>Palaeobiological Investigations: Research Design, Methods and Data Analysis.</u> eds. N.J.R. Fieller, D.D. Gilbertson & N.G.A. Ralph. Oxford: British Archaeological Reports (International Series), 266, 5-18.

ASHDOWN R. & EVANS C. (1981) Mammalian bones. In <u>Skeleton Green: a Late Iron Age and Romano-British Site</u>, ed. C. Partridge, London: Britannia Monograph Series 2, 205-35.

COLLIS J.R. (1970) Excavations at Owslebury, Hants: a second interim report. <u>Antiquaries Journal</u>, 50, 246-61.

COY J.P. (1987) Animal bones. In <u>A Banjo Enclosure in Micheldever Wood Hampshire</u>, ed. P.J. Fasham. Winchester: Hampshire Field Club & Archaeology Society Monograph 5, 45-53.

COY J.P. & BRADFIELD J. (n.d.) The animal bones from the Pre-Roman and Roman layers at Winchester, Western Suburbs, 1974-9. Unpublished Ancient Monuments Laboratory Report, 4063.

DONE G. (1986) The bones. In <u>Excavations on the Romano-British Small Town at Neatham, Hampshire, 1969-1979.</u> eds. M. Millett & D. Graham, Winchester: Farnham and District Museum Society/Hampshire Field Club and Archaeological Society Monography, 3, 141-147.

GRANT A. (1971) The animal bones. In <u>Excavations at Fishbourne 1961-1969</u>, ed. B. Cunliffe, London: Society of Antiquaries Report, 27, vol II 377-88.

GRANT A. (1975) The animal bones. In <u>Excavations at Portchester Castle (Vol. I).</u> ed. B. Cunliffe, London: Society of Antiquaries Report, 32, 378-408.

GRANT A. (1985) The animal bones. In <u>Guide to the Silchester Excavations: the Forum Basilica 1982-84</u>, ed. M. Fulford. Reading University: Department of Archaeology, 29-31.

JONES H. (1892) Note on the animal remains. In Excavations on the site of the Roman city of Silchester, ed. G.E. Fox. <u>Archaeologia</u>, 53, 285-8.

JONES R.T., LANGLEY P. & WALL S. (1985) The animal bones from the 1977 excavations. In <u>Excavations at Brancaster 1974 and 1977</u>, ed. J. Hinchcliffe with C.S. Green. Norfolk Archaeological Unit: East Anglian Archaeological Report, 23, 129-174.

LUFF R-M. (1982) <u>A Zooarchaeological Study of the Roman Northwestern Provinces</u>, Oxford: British Archaeological Reports (International Series), 137.

MALTBY J.M. (1981) The animal bones. In Excavations at Old Down Farm, Andover, part II: Prehistoric and Roman, ed. S.M. Davies. Proceedings of the Hampshire Field Club and Archaeological Society, 37, 81-163.

MALTBY J.M. (1983) The animal bones. In Excavations at Cowdery's Down Basingstoke, Hampshire, 1978-81, ed. M. Millett with S. James. Archaeological Journal, 140, 167-8, 176, 187-192, 258-9.

MALTBY J.M. (1984a) Animal bones and the Romano-British economy. In Animals and Archaeology (Vol.4): Husbandry in Europe, eds. C. Grigson & J. Clutton-Brock. Oxford: British Archaeological Reports (International Series), 227, 125-138.

MALTBY J.M. (1984b) The animal bones. In Silchester Defences 1974-80, ed. M. Fulford. London: Britannia Monograph Series 5, 199-212.

MALTBY J.M. (1984c) The animal bones. In The Romano-British settlement at Little Somborne. (ed. The Test Valley Archaeological Committee.) In The Southern Feeder: the Archaeology of a Gas Pipeline. Eds. P.D. Catherall, M. Barnett & H. McClean, British Gas Corporation, 136-41.

MALTBY J.M. (1985a) The animal bones. In The Prehistoric Settlement at Winnall Down, Winchester. ed. P.J. Fasham, Winchester: Trust for Wessex Archaeology/Hampshire Field Club Monograph 2, 25, 97-112, 137-8.

MALTBY J.M. (1985b) Assessing variations in Iron Age and Roman butchery practices: the need for quantification. In Palaeobiological Investigations Research Design, Methods and Data Analysis. Eds. N.J.R. Fieller, D.D. Gilbertson & N.G.A. Ralph. Oxford: British Archaeological Reports (International Series), 266, 19-32.

MALTBY J.M. (n.d.1) The animal bones from the later Roman Phases from Winchester Norther Suburbs: 1: The Unsieved Samples from Victoria Road Trenches X-XVI. Unpublished Ancient Monuments Laboratory Report, 125/87.

MALTBY J.M. (n.d.2) The animal bones from the Iron Age and Romano-British Phases of the Staple Gardens excavations, Winchester. Unpublished Ancient Monuments Laboratory Report, 4908.

MALTBY J.M. (n.d.3) The animal bones from the excavations at Owslebury, Hampshire: an Iron Age and Early Romano-British Settlement. Unpublished Ancient Monuments Laboratory Report, 6/87.

MALTBY J.M. (n.d.4.) The animal bones from the 1982-1983 excavations at Easton Lane Interchange (W29), Hampshire. Unpublished Ancient Monuments Laboratory Report, 7/87.

MALTBY J.M. (n.d.5) The animal bones from the 1978 excavations of the Late Iron Age and Romano-British Settlement at Abbotstone Down, near New Arlesford, Hampshire. Unpublished Ancient Monuments Laboratory Report, 58/86.

MALTBY J.M. (n.d.6) The animal bones from Brighton Hill South (Trenches B, C and K), Farleigh Wallop, Hampshire. Unpublished Ancient Monuments Laboratory Report, 155/87.

MALTBY J.M. (n.d.7) The animal bones from the 1973 excavations at Balksbury, Hampshire. Unpublished Ancient Monuments Laboratory Report, 4542.

MARPLES B.J. (1974) Report on the animal bones. In the Roman Fort at Longthorpe, eds. S.S. Frere & J.K. St. Joseph. Britannia, 5, 122-128.

MENSCH P.J.A. van (1974) A Roman soup-kitchen at Zwammerdam? Berichten van de Rijksdienst voor het Oudheidkundig Bodemonderzoek, 24, 159-66.

PARTRDIGE C. (1981) Skeleton Green: a Late Iron Age and Romano-British Site. London: Britannia Monograph Series, 2.

PFEIFFER J. (n.d.) The animal bones from the Prehistoric and early Roman phases of the Northern Suburbs excavations, Winchester.

THAWLEY C. (1982) The animal remains. In Early Roman Occupation at Cirencester, eds. J. Wacher & A. McWhirr, Cirencester: Cirencester Excavation Committee (Cirencester Excavations 1), 211-27.

BONE, ANTLER AND HORN INDUSTRIES IN THE URBAN CONTEXT

Arthur MacGregor

From the centuries immediately preceding the Norman Conquest of England, during which stirrings of urban development began to make themselves felt throughout northern Europe, there is ample evidence for an industry based on bone and antler that was already highly selective in the raw materials it used for particular purposes, coherent in its range of products and largely uniform in character over a very wide area. From the same period there are signs that horn too was being worked on a widespread basis, although the nature of the raw material (being much less durable under normal conditions of burial) presents problems in establishing the full range and character of the industry.

An attempt is made here to trace the respective fortunes of these crafts as urban settlements developed in size and number, a process that had an acute effect on the mechanisms by which raw materials were brought together and on the markets through which finished products were dispersed.

Items made from skeletal bone are often considered archaeologically along with those of antler. There are good reasons for this: antlers are merely bony outgrowths of the deer skull and when the distinguishing features of surface texture and internal cancellous structure have been removed they are difficult or impossible to tell apart. In any consideration of the industries which utilised these materials, however, it is of value to distinguish whenever possible between bone and antler, for when this is done several striking features become apparent.

EIGHTH TO ELEVENTH CENTURIES

Antler

In the period from about the 8th to the 11th century a noteworthy feature of this industry is the preference shown by manufacturers for antler rather than skeletal bone. In any society that relied heavily on deer meat for its sustenance such a preference would be instantly understandable, yet whenever comparisons are made between numbers of antlers and numbers of post-cranial deer bones recovered from early urban excavations, the latter invariably form only a fraction of the former. The paucity of deer bones and their under-representation compared with antler remains is a feature noted from numerous settlements of this period in northern Europe, including Lincoln (O'Connor 1982, 40), York (Kenward et al. 1978, 63; MacGregor 1982, 99) and Hedeby (Ulbricht 1978, 123). The implication is clear: the bulk of the antlers occurring in these earliest towns are to be seen not as food refuse but rather as raw materials for manufacturing.

From the antler remains themselves comes further evidence to support this claim. Even those sawn from the skulls of slaughtered animals are generally sufficient to outnumber the other skeletal elements from deer, but a consistently striking feature is the fact that most have been shed naturally in the wild and hence have deliberately been collected for utilisation. Such a pattern emerges wherever appropriate data have been gathered, as in the British Isles at York (MacGregor 1978, 46), Lincoln (Mann 1982, 44) and Dublin (O'Riordain 1971, 75), or on the Continent as at Arhus (Andersen et al. 1971, 121), Ribe (Ambrosiani 1981, 99), Hedeby (Reichstein 1969, 62-3), Menzlin (Schoknecht 1977, 68), Wolin (Müller-Using 1953, 64-7) and Dorestad (Prummel 1982, 119).

Only recently has an explanation been proposed to account for the development of such an apparently curious preference, in which readily available bone from food refuse or from the carcasses of traction animals was spurned in favour of antlers shed randomly and relatively inaccessibly over the countryside (MacGregor and Currey 1983). According to this proposition, the answer lies in the mechanical superiority displayed by antler over bone: various bending tests carried out on the two materials revealed that antler was consistently tougher than bone, the results varying by up to a factor of six times. To the manufacturer of combs, for example, in which the teeth are subjected to very considerable stresses, these differences would have been revealed all too clearly in the ability of their products to withstand wear-and-tear.

Accepting then, that the manufacturer had a sufficient appreciation of the superiority of antler to seek it out in preference to bone, how did he go about acquiring his raw materials? It seems inconceivable that any urban craftsman would have dedicated a great deal of time - even within the restricted shedding season - to searching out antlers scattered throughout the forests, thickets and moorlands of the countryside, where in any case the presence of foraging strangers would by no means have been welcomed by suspicious bailiffs. More probably the country-dwellers whose normal lives took them into those parts - foresters, herdsmen and the like - would have gathered up the antlers and stored them for subsequent sale to manufacturers. As for the antlers of slaughtered deer, it has been stated that few of these apparently reached the towns in association with carcasses and, even in the countryside, the numbers of individuals with legal access to deer was very limited. With the imposition of Norman rule, restrictions were further tightened and sanctions against illegal hunting were made more severe (see above). From at least this time and probably earlier, therefore, we may envisage that the bulk of the antlers reaching manufacturers from dead animals would have been channelled through the royal and noble households that operated a jealously guarded monopoly on hunting.

Although, such a conjecture seems plausible enough, the nature of the archaeological evidence from the British Isles is such that no corroboration has yet been forthcoming. In northern Poland, however, links have been adduced between the development of certain antler-working centres and the presence of noble households which controlled supplies of the necessary raw materials (Kurnatowska 1977, 124).

One important limiting factor in attempts to detect evidence of a trade in antlers is that all the settlements so far mentioned lie within the natural areas of distribution of red deer, from which the vast bulk of the raw materials derive. Nonetheless, some successful attempts at demonstrating such a trade have been made. By comparing burr sizes of antlers from slaughtered deer (presumed to be of comparatively local origin) with those naturally shed in the wild, Christophersen (1980, 156-9) was able to show with some degree of confidence that the material from early medieval Lund contained some imported elements. Similar success might be hoped for elsewhere, but the suggestion made here, that the flow of antlers from slaughtered beasts may also have been subject to a degree of control, leading, perhaps, to trading over no less a distance than that envisaged for shed antlers, would place a limit on the validity of Christophersen's distinguishing criteria.

Less equivocal evidence comes from regions lying close to established species limits. The occurrence, for example, of red deer antler among finished items and manufacturing debris at Birka, the earliest sizeable Viking age settlement in Sweden, implies deliberate importation since the site, on an island in Lake Mälar, is beyond the northerly limit for red deer at that period (Ahlén 1965, fig. 2). Elk rather than red deer formed the principal source of raw material for manufacturing at Birka and elk antlers naturally predominate there among waste materials (Ambrosiani 1981, 36). The occurrence of elk antlers at sites beyond the southerly limit of the species distribution once again points to an element of trade: Ribe and Hedeby have both produced evidence of this kind (Ambrosiani 1981, 52-3, 99), as has Dorestad in the Netherlands (Clason 1980, 246). It should be noted, however, that some doubt remains over the southerly limit of elk distribution in the early medieval period (MacGregor 1985, 37). Evidence for the utilisation of reindeer antler in the early Swedish towns of Lund and Lödöse (Ekman 1973, 49; Lepiksaar 1975, 234), both lying too far south for the antlers to have been found locally, also carries implications of long-distance trading.

In assessing the means by which antlers found their way into urban settlements, even within the ecological zones occupied by particular species, periodic visits by countryfolk to meets, courts and other regular gatherings no doubt played a part. It would be mistaken, however, to assume that manufacturers were necessarily passive or static during the early years of medieval urban development. No convincing manufacturing sites of this period indicating long-term settled production have yet been found. Even in the most productive Continental settlements such as Hedeby (Ulbricht 1978) it has long been recognised that the volume of production as represented by waste material from manufacturing is wholly inadequate as a basis from which to postulate full-time working. Until comparatively recent years this shortfall was generally explained by postulating some other trade which might have been practised on a part-time basis alongside antler working: on the evidence of what seem like rather fortuitous associations, amber working has been suggested at Menzlin (Schoknecht 1977, 69), Kolobrzeg (Cnotliwy 1956, 177) and Wolin (Cnotliwy 1958, 228), and stone working at Kalisz (Kurnatowski 1977, 123); alternatively, quite distinct activities such as hunting and farming were proposed at Hedeby (Ulbricht 1978, 122). Recently, however, the view has been gaining ground that this dearth of remains from full-time activity should be explained not by postulating that any one craftsman had two or more jobs to perform, but by assuming that he followed only one trade in more than one place: in other words, before the establishment of settlements sufficiently populous to support full-time producers, the majority of antler workers were itinerant. Such a conclusions has been reached independently by Kurnatowska (1977, 122), Ambrosiani (1981, passim) and Christopherson (1980, 164-5), and has a great deal in its favour, not only in reconciling the seemingly conflicting evidence for 'professional' production with the absence of long-term workshops but also in explaining the widespread uniformity of types, promoted by the mobility of crafsman. Needless to say, the territory covered by any one worker would have been limited in extent, but the opportunity for overlap - particularly at the seasonal gatherings already mentioned - would have ensured widespread distribution of common practices and motifs. With this hypothesis we can begin to comprehend, for example, the striking uniformity remarked upon by Ambrosiani (1981, 161) amongst combs from settlements stretching from Staraja Ladoga in the east to Dublin in the west.

As well as introducing the manufacturer to a wide spectrum of customers, this method of operation would have had the advantage of bringing the producer into more direct contact with the suppliers of his raw materials - the peasants and others who would most readily find the antlers shed in the wild, and, perhaps, the noble households in which the antlers of slaughtered animals would have accumulated.

This, then, is the picture that seems to have most to recommend it in what might be called the proto-urban phase: the antler worker, a manufacturer principally of combs and a few other items such as playing pieces, dice and handles, a somewhat transitory figure in the urban scene, absent for extended periods while he toured the countryside, selling his wares here, acquiring raw materials by purchase or exchange there. He worked in a single material, which he exploited exhaustively.

Bone

As to articles made of bone, it is striking that the bulk of these clearly involved no significant level of craftsmanship and were evidently made by the individual as required. A great proportion of the simpler bone pins can be consigned to this category, as can many of the craft tools such as pin beaters and spindle whorls made from femur heads as well as everyday items such as whistles, skates, toggles and the like. This leaves a residue of more carefully made items, which may or may not have been professionally manufactured, a few such as composite combs which almost certainly were, and a small amount of undeniable industrial waste.

As already mentioned, antler is a measurably superior material for making items like combs which have to endure considerable stress, but a certain proportion of combs was nonetheless made of bone. Sometimes the connecting plates would be of split ribs or other bones, while the tooth-plates which took the real stress would be of antler. At other times, inexplicably, this situation would be reversed, and in some instances all the elements were of bone. As yet there seems to be no evidence in Britain for composite comb makers who relied entirely on bone, and it may be that it was used merely when antler was scarce. At least one production site has been found on the Continent, however, which was supplied exclusively by horse bones: excavated at Münster in the 1970s, this site produced evidence for a comb-making industry in the 8th century that relied wholly on bone rather than antler: blanks, roughouts and finished combs were all found on the site (figure 1) as well as some 300 sawn ends of horse metapodia discarded in the manufacturing process (Winkelmann 1977, 111-115).

Although no contemporary site matches the scale of Münster, scattered evidence indicates that similar work practices were followed in towns over a wide area of northern Europe. Sawn epiphyseal ends from cattle metapodials similar to those of horse discarded during comb making at Münster occur at Dorestad (Clason 1978, 294-5; Prummel 1983, 259) and in some numbers at Southampton : here seven Late Saxon pits were filled with offcuts of this type, though without evidence as to the nature of the end products (Holdsworth 1976, 45). Such a geographically widespread uniformity of

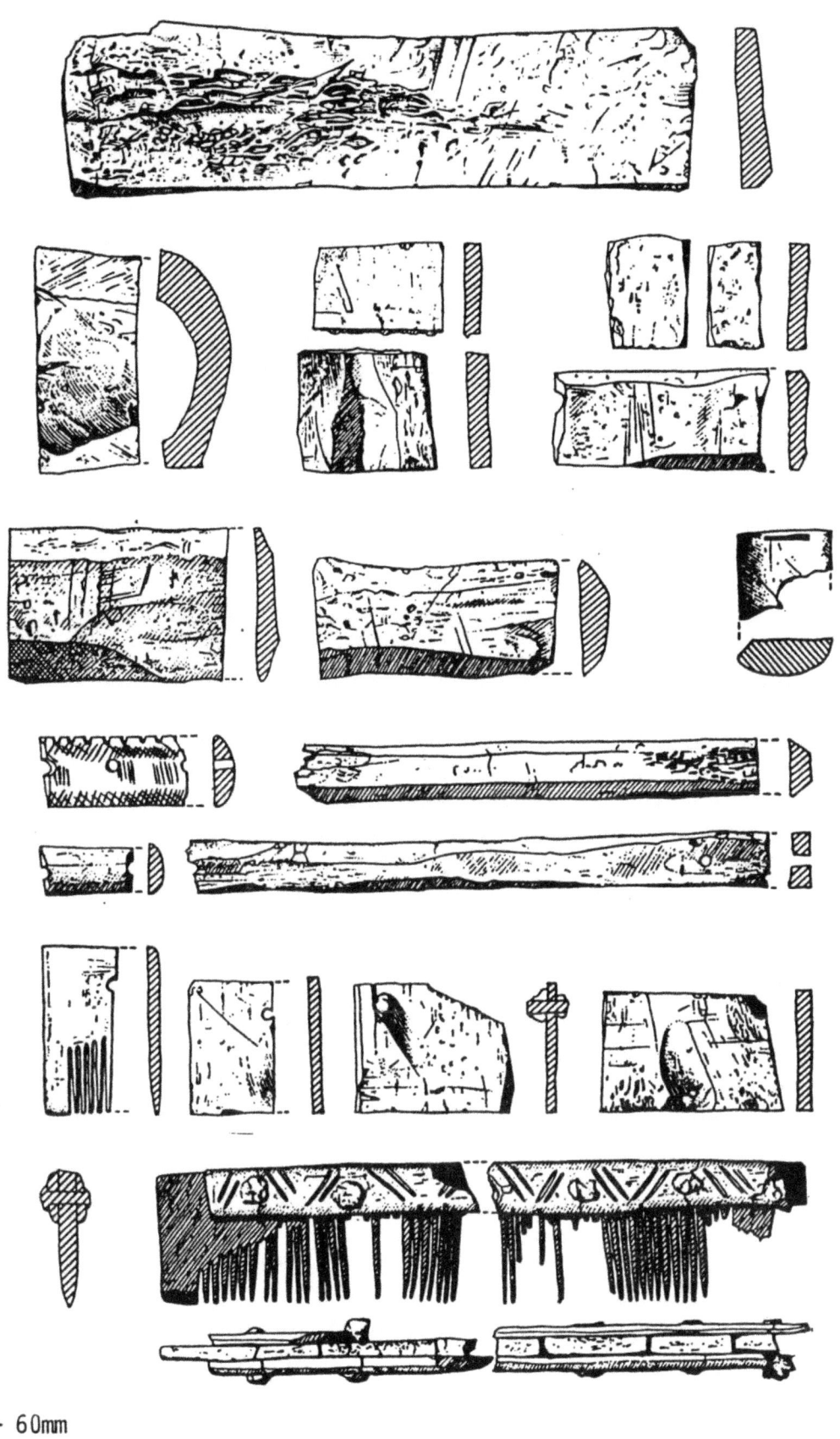

Figure 1. Manufacturing debris and finished articles from the production of bone combs in 8th-century Münster. After W. Winkelmann (1977).

practice speaks of a well-established craft with a consistent approach to working routines, so that even if the evidence is more sparse than that for antler working, it serves to show that by the 8th or 9th century there were numerous craftsmen working routinely in bone.

That bones other than metapodials were utilised on a systematic basis is shown by comb-making waste derived from cattle mandibles found at Dorestad (Prummel 1983, 259) and by the contents of a Viking age pit at Leadmill Lane in York, which included (along with antler shavings and offcuts) many fragments of split ribs from the manufacture or repair of combs or comb-cases (MacGregor 1982, 94-5). A larger concentration of ribs, amounting to some 600 pieces, was found in Late Saxon levels at Thetford, where, however, their function was less apparent (Rogerson and Dallas 1984, 199).

Horn

Any assessment of the pre-Conquest horn industry is made that much more difficult by the infrequency with which horn survives in archaeological contexts. From our knowledge of medieval and later horn working, in which a major feature was the use of pits in which horn was soaked for a period of weeks before working, it is difficult to see how an itinerant phase could have preceded a settled one in this craft. (For further discussion of this question, see below, p. 117). More importantly, it is hard to imagine that before the development of sizeable towns, attracting a steady flow of goats, sheep and, especially cattle, there could have existed a mechanism by which the raw materials for a settled industry could have been brought together.

The possibility that extensive use as drinking vessels could have been made of entire horns, requiring little more than trimming around the base when they had been pulled from the skull, seems reasonable enough (MacGregor 1985, 151-3). The horn handles of which evidence survives in the form of corrosion products on large numbers of Anglo-Saxon knives and swords (see, for example, Watson 1984), could have been cut from the solid tips of horns without any pretreatment and even scale handles may have been produced in this way. It would be wrong, however, to imagine that Anglo-Saxon craftsmen were not already familiar with the possibilities of flattening or bending horn with the aid of heat: the Benty Grange helmet, dating to the 7th century, has been judged to have been fitted with horn plates treated in this way (Bruce-Mitford and Luscombe 1974). A series of 9th- or 10th-century horn combs fitted with bone side plates clearly involves the use of flattened horn (MacGregor 1985, 95); later horn combs have no such strengthening strips and it may be that their use here implies a phase when techniques were still far from perfect and when inexpertly-flattened horn had to be held in shape with splints.

Even in the absence of finished artefacts the development of horn industries can be detected in some of the earliest towns. At Dorestad, for example, the presence of many sawn horncores of sheep, goat and cattle testify to the presence of an intensive industry (Clason 1980, 239). At Ribe the numbers of cores found exceeded those for post-cranial bones of the same species, indicating an industrial interest, while the manner in which some of them had been cut up (apparently, with the sheath still attached to the core) suggested that comb-making accounted for at least some of this activity (Ambrosiani 1981, 100). Horn cores with similar cut marks were found at Menzlin (Schoknecht 1977, 123, Taf. 31). No evidence was found at any of these sites to suggest that horn working was practised in direct association with either bone or antler working, although the horn combs with bone side-plates mentioned above indicate some degree of overlap of interest.

ELEVENTH TO THIRTEENTH CENTURIES

In the course of the 11th to the 13th centuries, which saw continuing urban aggregation and a tightening of legislation governing the countryside, far-reaching changes were wrought on the situation just described. If antler workers were indeed itinerant in earlier centuries, there is now some evidence for the movement of workers into a more sedentary mode of existence. For example, at Wolin, Cnotliwy (1958, 228) cites the discovery of a workshop established in a reed hut over several generations up to the 12th century.

A more general feature, however, is a marked decline in the utilisation of antler, compensated by a rise in the use of bone, and, more particularly, of horn. There was also, it may be suggested, a radical realignment of the basis on which the trades utilising these materials were organised.

Comb manufacture

To return again to combs, which form the most useful indicators in tracing the course of these events, this period saw first of all a shift towards the utilisation of bone, rather than antler for composite combs, most vividly demonstrated at Schleswig, the early medieval settlement that replaced Viking-age Hedeby (Ulbricht 1984, 43, 46-7), followed by the decline and extinction of the composite comb and its replacement by smaller, lighter combs of one-piece construction, again normally of bone rather than antler. As the mechanical advantages enjoyed by antler in earlier centuries continued to apply, the reasons for its being abandoned at this time are not immediately clear. There was no marked decline in the red deer population, although the extension of forest laws may have made access more limited. In early medieval Schleswig, Ulbricht (1984, 73) has noted a shift in the ratio of antlers utilised, away from shed specimens and in favour of slaughtered animals, perhaps indicating a tightening of control by noble families on hunted game.

Kristina Ambrosiani, in her review of Scandinavian comb making (1981, 162), considers the possibility that legislation controlling hunting may have been extended there; stressing the continuing importance in Scandinavia of shed antler, however, she quotes Olaus Magnus writing in 1555 as authority that 'the antler which [deer] shed in the forest ..... can be taken by those who find it.'

In England game laws were certainly extended under William I and his successors, while penalties for offences against them reached a new level of harshness: blinding was the ultimate punishment (Anglo-Saxon Chronicle, sub anno 1087).

That the value of antler as well as venison came to be appreciated in England is demonstrated by an order recorded in 1224-25 in the Close Rolls of Henry III, instructing one Hasculf de Adhelkeston to 'make over all the antler beams which he has at his disposal from our forest which is under his care to Philip Convers the crossbow-maker for the manufacture of crossbow nuts' (Rotuli Litterarum Clausarum 2,50). Whether Hasculf had amassed his collection by systematically gathering specimens shed in his forest or whether he had acquired a store of antlers from deer killed in the exercise of hunting rights is unrecorded, but either way there is an implication that access to antlers was, at least in certain times and places, subject to controls.

Two further factors may have had a hand in the trend away from antler: firstly, as urban populations expanded to the point where increasing numbers of craftsmen could be supported on a sedentary basis, direct links with the country-dwellers through whom fresh antlers had been channelled were broken; secondly, the bone refuse generated by urban populations may simply have become too voluminous to ignore. Any reservations that may have been held concerning the inferior performance of this material were overcome by modifying the design of combs: from about the 11th century they take on a markedly lozenge-shaped cross-section, resulting in blade-like teeth that compensated for their lower bending strength with increased breadth (see MacGregor 1985, 81).

Other changes were also afoot at this time. Whereas the antler worker had enjoyed a virtual monopoly of comb making in Anglo-Scandinavian England, a variety of materials came to be used for this purpose in the early medieval period: as well as the lozenge-sectioned one-piece bone combs already mentioned, medieval excavations produce others in wood – mostly boxwood – of identical design. It seems inconceivable that two sets of craftsmen could have existed in parallel – the boxwood comb makers and the bone comb makers – and surely we must be looking here at the products of single group – comb-makers working in whatever medium was required of them.

As time went by combs in both these materials tended to become larger and flatter, seemingly under the influence of horn which may yet have been worked in the same workshop. By this time no bone strengthening strips were needed on horn combs, and the lightness and toughness of horn would have made these distinctly preferable to combs in other media. What is implied in this new situation is a complete realignment of the industry from a material-based craft, to a product-based industry in which a single craftsman – or by this time, a single workshop – might produce only combs, but in a variety of materials according to the tastes of the consumer and the resources of the maker.

Other objects

A similar shift can be postulated for other parts of the industry in which antler workers formerly held a major stake. From the records of London's medieval Cutlers' Company, for example, it appears that from their earliest days their numbers included not only blade-smiths but also a distinct group of hafters who made handles of every material, combining work in bone or horn with metalwork and even with precious metal according to demand (Welch 1916-23, 19-21). The successful operation of their monopoly depended not on exploitation of a single material but on their ability to combine and work a range of substances. In any attempt to evaluate the medieval bone industry in the capital, therefore, the numerous well-made handles found there which might seem to be indicators of a rather successful industry are more likely to be testaments of its demise in the face of advances by the cutlers in extending their control over every part of their operations.*

As yet the evidence from other towns is not very great, but it is significant in this context that excavations in Worcester produced evidence that bone and perhaps horn were being worked in a single site along with iron and bronze, apparently on a workshop producing knives (Carver 1980, 174-8).

Again a similar move in favour of products and away from materials can be detected among playing pieces from medieval excavations, although in this case there seems to have been no single craft guild exercising a monopoly on their production. Antler continues to be used here, being particularly suited to this purpose because of its solid structure, at least towards the base. Bone is used almost as frequently for tablemen, which were discoid pieces cut generally from cattle mandibles or from deer pedicles, though in the recently discovered set from Gloucester it seems that bone from deer skulls as well as pedicles was used (Stewart and Watkins 1984). Bone was also used for chessmen, even though it is singularly unsuited for this purpose since the cylindrical long-bone shafts have to be plugged with solid tissue before use (MacGregor 1985, 138, fig. 73).

Other items commonly made from bone in the medieval period include buttons and beads. Characteristic waste from these products has been found at King's Lynn (Clarke and Carter 1977, fig. 143, 25-8), Hull (Armstrong 1977, 70) and elsewhere, although only in Coventry did they occur in significant concentrations (Gooder et al. 1964, 129). There is little indication, however, that the makers of buttons produced any other lines, further evidence that the concept of a cohesive bone industry is one that cannot be sustained for medieval England. While it is at times difficult to decide whether particular waste pieces or even finished products are buttons or beads, important finds from Saint-Denis present no problems of interpretation (figure 3). These were confidently interpreted by the excavator as refuse from the manufacture of rosaries (Meyer 1979, 2.2.1). As a popular centre of pilgrimage Saint-Denis would have been well-placed to support such a producer, and it would be reasonable to envisage other such manufacturers as part of the industry dedicated to the production of religious souvenirs in other towns such as Canterbury.

Hornworking

The centuries after the Conquest were a period of growth for the horners, unlike the antler workers, and the unity of their craft was recognised by the formation of professional guilds in London and York. Even in those areas such as knife-hafting and comb-making in which the horners exercised no monopoly over production, they were, it may be suggested, firmly in charge of the supply of raw materials.

The most impressive illustrations of the medieval horn-worker's practice

---

* Note: an extensive account of knives from excavations in London, together with related evidence for their production, has been published since this paper was compiled (Cowgill et al. 1987); figure 2 shows 14th/15-century examples from that publication.

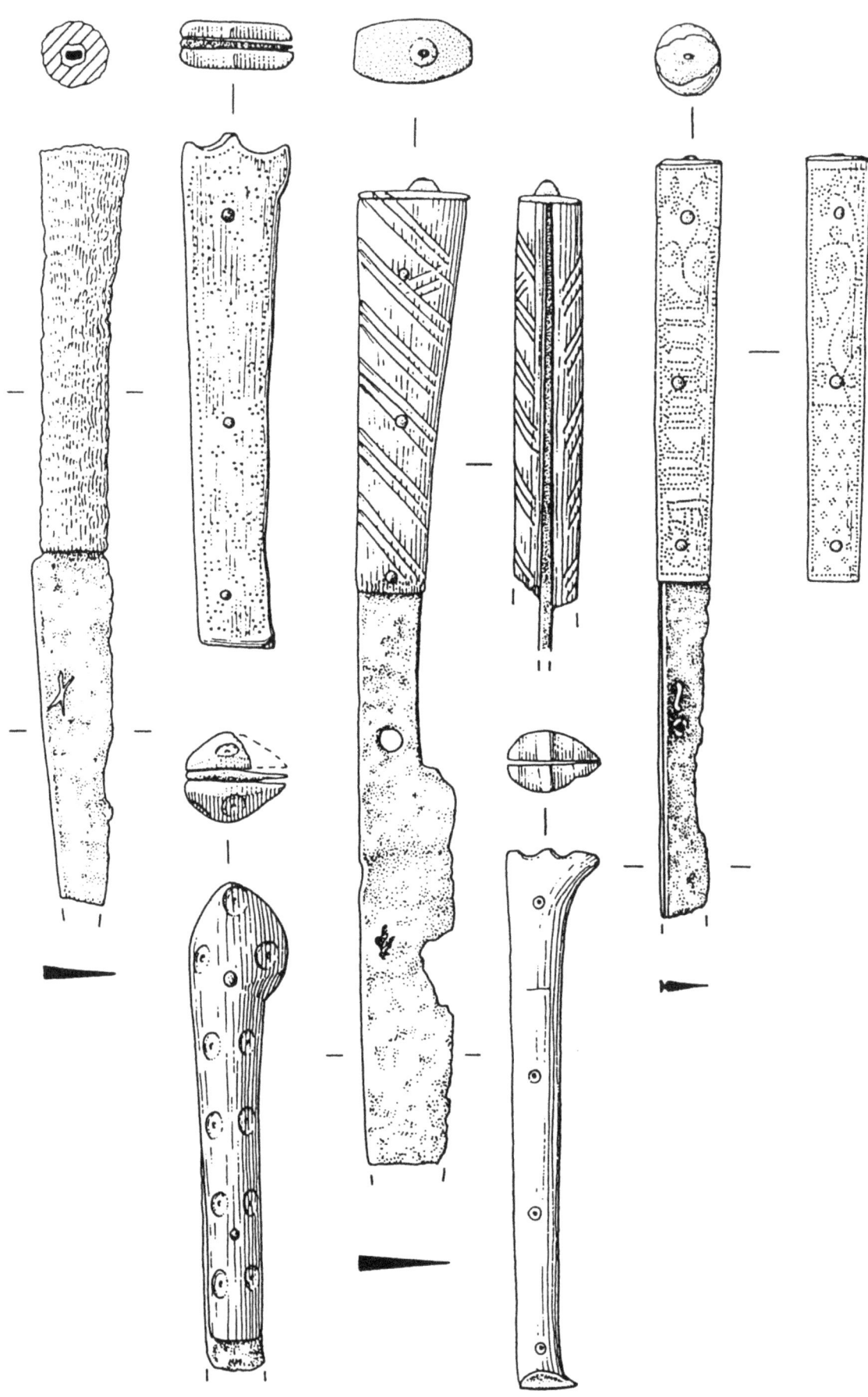

Figure 2. Knife handles of the 14th/15th century from London: the first is of horn, the remainder of bone. After J. Cowgill et al. (1987). Scale 1:1.

comes from Hornpot Lane in York, excavated in 1957-8 (Wenham 1964). Here a shallow clay- and timber-lined pit containing some 200 horn cores - mostly of cattle and goat - was uncovered, along with a series of hearths. The pit corresponds with the soaking pit used by English horners until recent times for steeping horns to loosen the sheaths from their cores and, seemingly, to make them more malleable. Quoting a worker recalling the horn industry at the beginning of the present century, Wenham (1964, 51) gives six such pits as the minimum viable number for a flourishing workshop and two or three months as the average soaking time.

This element of the horner's craft practice is of some importance in assessing the early development of the industry, for soaking horns would seem to be irreconcilable with an itinerant way of life. Not all authorities are agreed, however, on the need for prolonged soaking. At Southampton, Bourdillon and Coy (1980, 97) noted an absence of cut marks around the bases of the horn cores and speculated that the horns might have been imported from some distance, during which process the bond between sheath and core had time to rot, so that the sheath could simply be pulled off. According to an 18th-century source quoted by Prummel (1978, 409) this was indeed the normal practice: 'the horns are left lying in the open air, until they begin to decompose, when the horn can easily be separated from the bony part.' Even up to the present century, it seems that horns were treated in this way by German horners (Andés 1925, 43). Perhaps it should be assumed, therefore, that the lengthy soaking favoured by English horners had more to do with improving the working qualities of the horn than with separating the sheath from the core, in which case the horner need not necessarily have been tied to a single workplace. The problem is an interesting one in the history of urban crafts, but seems insoluble for the present.

Whichever method of separation was used, the solid tips of the sheaths were sometimes cut off in advance, perhaps with a view to hastening the separation of the sheath from the core but also, perhaps, because the tips could be used without further treatment for the manufacture of handles, buttons and the like. In some instances the sheath was cut into cylindrical sections while still on the core, so that the cores too are sometimes found divided into sections. At other times saw cuts are found barely marking the surface of the core, presumably indicating that the sheath was cut through and removed after some preliminary rotting of the natural bonding.

Some of the resulting horn cylinders were turned directly into beakers, inkwells, and the like, while others were subjected to further treatment involving prolonged boiling, slitting up one side and flattening under pressure into leaves or sheets (MacGregor 1985, 66-7).

No hearths for this process have been recognised other than at Hornpot Lane, but concentrations of horn cores are not uncommon and are readily interpreted as evidence for horn-working. At Angel Court in London, for example, fifty horn cores of cattle and fifteen of sheep were found in a single medieval layer, many of them sawn into sections (Clutton-Brock and Armitage 1977, 88, 93); cattle, sheep and goat horn cores were also represented, or rather over-represented at Baynards Castle (Armitage 1977). Other concentrations have been found in Bristol, Exeter, Hereford, Oxford, Coventry, King's Lynn, Northampton, Newcastle, Perth and York. At Northampton fifty-four cattle horn cores were found in a pit containing only twelve other cattle bones while 164 horn cores came from an adjacent pit; it

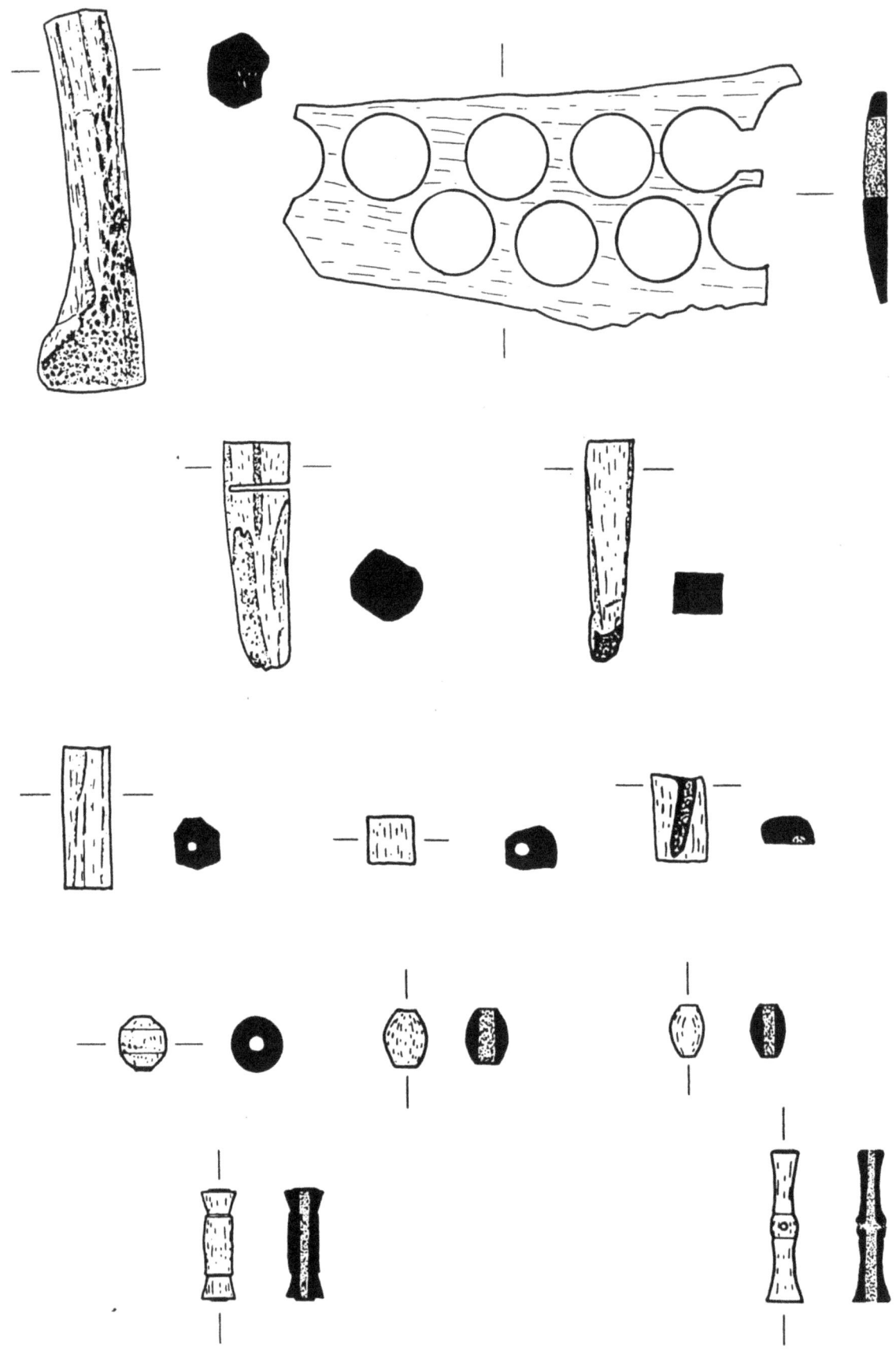

Figure 3. Waste material and finished products from the manufacture of bone rosary beads in late 15th/16th century Saint-Denis. After O. Meyer (1979). Scale 1:1.

was noted too that part of the skull was normally removed with the horn cores, presumably with the intention of preserving as much of the horn as possible (Harman 1979, 328, 331). At Skeldergate in York a few horn cores similarly treated and others chopped through at the base were found in a series of 12th-century pits (O'Connor 1984, 20, 28-9). Here there were not only abundant cattle horn cores but also fifty examples from goat, a species notably absent among the other skeletal remains from the site. Comparison of skeletal fragments and horn cores suggested that in the early post-Conquest phases horns were being salvaged from the remains of slaughtering and butchery; then in the course of the 12th century increasing importance was attached to this activity to the extent that horns from goats were actually being imported to the site, presumably to supply the needs of an established horn industry. Elsewhere a 13th-century tenement at Perth has produced some 1500 horn cores (Bogdan and Wordsworth 1978, 27), while in Lincoln the remains of some 1100 cores were recovered from a stone-lined pit of c. 16th-century date (Jenny Mann, pers. comm.). Continental evidence in the form of horns thought to have been assembled for working comes from early medieval Lund (Ekman 1973, 67).

Not all such concentrations of horn cores, however, should necessarily be interpreted as indicating the presence of workshops nearby. The wide scatter of these finds up and down the country indicates that horns were salvaged for working on a fairly wide scale. Horning would almost certainly have been an exclusively urban craft, however, since it relied on a certain volume of raw materials only likely to be satisfied in this context. On the other hand, almost any slaughterer might have been induced to save up horns for sale to a horner working at some distance away. Consistent removal of horn cores from sheep skulls has been noted, for example, at Ospringe in Kent (Wall 1981, 239), which might seem an unlikely rural setting for a horn industry. Some historical support for the idea of a trade in untreated horn is supplied by an act of Parliament in 1465 that prohibited the export of rough horns from within a 24-mile radius of London until the needs of the London Horners' Company had been satisfied (Fisher 1936, 39). Further references to trade in horns have been assembled in Appendix 1.

Another cause for caution in interpreting these concentrations of horn cores is raised by the possibility that in some instances they may indicate the activities of tanners rather than horners (Serjeantson, this volume). A series of 13th-century pits in 's-Hertogenbosch in the Netherlands associated with some 76 horn cores of goat and 54 of cattle, have been published as tanning pits by Prummel (1978, passim), who notes that half the horn cores from 's-Hertogenbosch had a hole driven into them after death, suggesting that the carcass may have been hung up by the horns on a hook to make handling easier. This feature was also noted on many of the horns from Perth. Driving a hole through the horns would have made them unacceptable to the horners: perhaps the supply of horns from food animals at times so far exceeded the demand that some were simply discarded. This being so, the business relationship that might reasonably by postulated between tanner and horner must at least have had periods when the resources of one outstripped the other, and this may account for some of the dumps of horn cores on tannery sites. The potential importance of tanners in supplying the horn industry is demonstrated in documentary evidence from 18th-century Norfolk, where control of the livestock trade is said to have been under the control of the tanners whose primary concern was for the skins rather than the flesh (Noddle 1975, 257).

What then were the principal methods and products of this horn industry? The processes in the eighteenth century are illustrated by Diderot (figure 4). Mention has already been made of horn beakers, consisting of a cylindrical section of horn, forced over a former (if we are to assume that Diderot's description of 18th-century French practice can be projected into the past) in order to eliminate the natural curvature, and grooved around the inside of the wall to receive a base in the form of a disc of horn or other material. These probably formed an important part of the horner's repertoire up to the Tudor period, when the rise of glass began to make inroads into their trade. Horn inkwells were made in a similar manner, and horn spoons which also demanded moulding under heat at some point in their manufacture were no doubt also a popular line. By the 17th century, however, we get our first glimpse in the historical record of what must also have been a major feature of the medieval horn industry. At this time, it was said, over two million leaves of horn were being exported from London (Fisher 1936, 14): that is to say, what must have been a large proportion of the horn passing through the horner's hands was simply being processed into flat sheets before being sold on to the other craftsmen for working into finished products. The greater part of the Horners Company must therefore have been made up of what were known in the later parlance of the trade as horn pressers or horn breakers. Opening the horn and pressing it into sheets must always have been regarded as central to the horner's craft and even in the late 17th century the London horners were still vigorously defending this aspect of the trade from encroachment by the comb-makers (Fisher, 1936, 6). The relationship between them seems likely to have been a long standing one, stretching back to the early medieval period when the comb-makers would have been buying flattened sheets of horn to work in the same way as wood and bone. Raw materials would have been bought in exactly the same way by the cutlers - solid tips for tanged blades, flattened plates to make the scales for strip-tangs. Needless to say, such a degree of specialisation would have been much less of a feature in the smaller provincial towns: evidence for all stages of horn working within single institutions, ranging from opening or breaking the raw horns to pressing them flat and working them into combs and lantern leaves comes from 18th- and 19th-century Kenilworth (Drew 1965, 22).

By 1692 the London horners acknowledged that their trade chiefly depended on making lantern leaves. Again these would have formed a mainstay of the medieval industry, providing panes for the doors of simple cylindrical lanterns, though demand was increased appreciably by the adoption in the 17th century of new patterns of lanterns with multiple panes. After opening and flattening a suitably light-coloured horn in the normal way, lantern leaves were produced by delaminating or splitting the horn into thin sheets, which were then further greased between well-greased and heated iron plates; the translucency of the plates could be further improved by smearing them with oil and warming them over a fire (MacGregor 1985, 67; Poller 1980, 124-5). Two fragments of pressed horn from a 14th-century layer at Baynards Castle have been tentatively identified as lantern leaves by Armitage (1977). Similar leaves were evidently in demand as window panes during the medieval period. I know of no instance of horn panes having been recognised in medieval buildings but we know of the practice from a lament of 1580, after glass panes had ruined the trade, to the effect that 'horn in windows is now quite laid down' (Fisher 1936, 13).

Bone and antler

Few new lines developed in the post-medieval period to compensate for such losses. Antler found only occasional use as a raw material: the imbalance of supply over demand is graphically illustrated in the manner in which antlers came to be prized principally as hunting trophies; increasingly efficient methods of killing deer with firearms were paralleled by a collapse in the industrial demand for antler. Bone survived into the era of mass production for the manufacture of buttons and combs and found a new market in the expanding demand for brushes, especially tooth-brushes, handles for which were made by the thousand in bone up to the present century (Woodall 1959). Horn too enjoyed a continuing appeal for the making of combs (especially elaborate combs to be worn in the hair) and became the principal material used in the manufacture of powder horns for both military and civilian use.

A rare documentary and pictorial record from the late-seventeenth-century Germany confirms the picture of fragmentation that has been outlined above (Beigel 1698, 452-9); the crafts mentioned in this source include comb-making (figure 5a), in which boxwood, elephant ivory, tortoiseshell and walrus tusks are all said to have been worked by the same craftsmen, and rosary-making (figure 5b) in which work in bone was combined with wood, amber, coral, hardstones, gems and metal. (The rosary-makers, incidentally, who would cover them with silk or decorative threadwork, lending weight to the present writer's suggestion that the bone discs commonly found on medieval excavations may have been merely the 'skeletons' of covered buttons).

Other outlets developed for items such as spectacle frames, snuff mulls and cigarette holders. For a period in the nineteenth century, greater possibilities were glimpsed by the industry with the development of techniques of reconstituting powdered horn in heated moulds to produce items of complex structure such as electrical switch covers. Ultimately, however, this proved to be not the dawn of a new era of expansion for horn products but rather the prelude to their extinction, for it was found that the same technology could be applied with greater facility to various man-made materials, the forerunners of modern plastics.

CONCLUSION

To sum up, it may be suggested that the features distinguishing industries based on bone, antler and horn in the medieval and post-medieval periods from those which had exploited these materials in the Anglo-Saxon and Viking periods were closely bound up with the rise of urbanism. The development of increasingly large and increasingly numerous centres of population provided opportunities for the settled practice of these crafts. With growing numbers of craftsmen adandoning their traditionally itinerant way of life for the benefits of sedentarism, direct links with the scattered rural populations that had supplied a large part of the raw material for the antler industry were gradually broken; ever stricter controls exerted over these materials may also have interfered with supplies. By way of replacement, much more importance was accorded to the by-products of the trade in animals that kept the town supplied with food and with means of traction. The inferior performance of bone for this source was compensated for by the evolution of new designs, notably for combs. Once a demand had been established, we may envisage the systematic salvaging of specific bones

Figure 4. Interior (somewhat idealised) of a horner's workshop, from Diderot's Encyclopédie (1771). The processes illustrated include the warming of horn at the bench (a) and on the hearth (b); cutting (c); opening (d); flattening with the aid of a wedge press (e) and a screw press (f); moulding on a former (g); and cutting rough-outs for further treatment (h).

a   b

Figure 5. Workshops of (a) comb-maker and (b) rosary-maker from Weigel 1698.

from the slaughterhouse for selling on to the trade. The horns of cattle, sheep and goats would similarly have been reserved for manufacture, and the supply of raw materials for the tanners and the horn workers was no doubt closely bound together. In responding to the changing conditions of urbanised society, the manufacturers who worked in skeletal materials had to adopt new working practices and new craft allegiances. The seemingly unbroken sequence formed by their products in the archaeological record disguises a series of upheavals that were no doubt as painful as they were far reaching.

Appendix

In addition to the sources cited in the text, the following references shed some light on the trade in horns and horn products. I am grateful to Mr Nigel Ramsay for drawing my attention to the London references.

Customs Accounts, London (Public Record Office)

E.122/71/13 1390 (tonnage and poundage)

m. 8^V Roger Crane: spectacles, ivory, etc.

m. 9 Jo. Otteley: spectacles, swords, etc.

m. 13^V Peter van Crowemer': 21b. ivory (17s. 7d.), 9 doz.

  spectacul' (6s.), 2200 cornubus pro lucernis (14s. 8d.), etc.

m. 14^V Gregory Hardbon: incl. 3500 cornubus pro lucernis (28s.)

m. 27^V Jo. Andrew: incl. 800 cornubus bovium (32s.)

Note. Amongst the material listed here it is surprising to find in addition to expected imports (ivory, spectacles) not only raw horns (cornubus bovium), in which England might have been expected to be self-sufficient, but also panes for lanterns (cornubus pro lucernis). Clearly the industry which resulted in a profitable export trade in the sixteenth century was as yet insufficiently developed to supply the entire needs of the home market.

E.122/73/10 1438- (petty custom)

fo. 12^V (?on a galley) 4 gross of glas., 5 gross of spectacle cases, 2 gross of penners & inkhorns, etc.

E.122/77/4 1442-3 (tonnage and poundage)

m. 6   20 dozen spectacle cases

m. 11^V 4 gross of inkhorns, & ditto of pens

m. 27^V 17 blowyng hornes; 7 goblets de cornibus; etc.

E.122/73/20A 1446 (import ledger)

fo. 4^v (on a ship from Danzig); 5 dozen spectacle cases; 6 dozen pens & inkhorns

fo. 20^v (ship from Antwerp) 1 gross of spectacles

E. 122/73/25 1450 (tonnage and poundage)

fo. 31 5 doz. drynkynghornes

E.122/79/5 1494- (imports and exports)

m.1^v or 2 Henry Hill: 4 gross of spectacles; etc.

m.4^v Ric. Harryson: 6 dozen leatherbottles & 7 dozen pouchenyng'; etc.

m.8^v Barth. Mowger: incl. 4000 tennes ball, 300 lettirbox, 12 gross of spectacles

m.12 Andrea Berion: incl. 1 gross of penners & inkhorns

Customs Accounts, Hull (Childs 1986)
Imports

61 Navis Hans Wynrik vocata Mare de Dansk applicuit 24 die Augusti 1463

    60 scok tabylmen (1 scok = 60 pieces)

    20 par tabylles (par = pair or set)

145 Navis Michaelis Johnson vocata Jacob de Westerscowe applicuit

    19 die Juni 1471

    60 inkhorns

211 Navis Johannis Mensen vocata Cristofer de Hamsterdam applicuit

    ultimo die Marchii 1489-90

    5 croftes whytt hornes (croft = ?basket)

Exports

44 Navis Johannis Person vocata Jacob de Fere exivit 7 die Marcii 1463

    3 m cornuum bovinorum

90 Navis Corneli de Grave vocata Mary de Hull exivit 20 die Maii 1465

    8 c cornubus bovinis

Acknowledgements

For permission to reproduce illustrations I am grateful to Professor Wilhelm Winkelmann and the Altertumskommission für Westfalen (Fig. 1); to Jane Cowgill, Nick Griffiths and the Museum of London (Fig. 2); and to Olivier and Nicole Meyer and the Unité d'Archéologie de Saint-Denis (Fig. 3).

REFERENCES

AHLEN, I. (1965) Studies on the red deer, Cervus elaphus L., in Scandinavia I: history of distribution. Viltrevy, 3, 1-88.

AMBROSIANI, K. (1981) Viking Age Combs, Comb Making and Comb Makers in the Light of Finds from Birka and Ribe (Stockholm Studies in Archaeology 2) Stockholm.

ANDERSEN, H.H., CRABB, P.J. and MADSEN, H.J. (1971) Arhus Søndervold: en Byarkaeologisk Undersøkelse (Jysk Arkaeologisk Selskabs Skrifter 9) Copenhagen.

ANDES, L.E. (1925) Bearbeitung des Horns, Elfenbeins, Schildplatts der Knochen und Perlmutter Leipzig and Vienna: Hartleben.

ANGLO-SAXON CHRONICLE (1961) The Anglo-Saxon Chronicle, ed. D. Whitelock. London: Eyre & Spottiswoode.

ARMITAGE, P.L. (1977) The Mammalian Remains from the Tudor Site of Baynards Castle, London: a Biometrical and Historical Analysis (Ph D thesis, University of London).

ARMSTRONG, P. (1977) Excavations in Sewer Lane, Hull, 1974. East Riding Archaeologist, 3, 1-77.

BERGQUIST, H. and LEPIKSAAR, J. (1957) Animal Skeletal Remains from Medieval Lund (The Archaeology of Lund: Studies in the Lund Excavation Material 1) Lund.

BOGDAN, N.Q. and WORDSWORTH, J.W. (1978). The Medieval Excavations at the High Street, Perth, 1975-6 Perth: Perth High Street Archaeological Excavation Committee.

BOURDILLON, J. and COY, J. (1980) The animal bones. In P. Holdsworth, Excavations at Melbourne Street, Southampton, 1971-76 (Council for British Archaeology Research Report 33) London, 79-118.

BRUCE-MITFORD, R. and LUSCOMBE, M.R. (1974) The Benty Grange helmet. In R. Bruce-Mitford, Aspects of Anglo-Saxon Archaeology, London: Gollancz, 223-42.

CHILDS, W.R. (ed.) (1984) The Customs Accounts of Hull 1453-1490 (Yorkshire Archaeological Society Record Series CXLIV) Leeds.

CARVER, M.O.H. (1980) *Medieval Worcester: an Archaeological Framework* (Transactions of the Worcestershire Archaeological Society 3rd series 7) Worcester.

CHRISTOPHERSEN, A. (1980) *Raw Material, Resources and Production Capacity in Early Medieval Lund* (Meddelanden fra̋n Lunds Universitets Historiska Museum new ser 3), 150-65.

CLARKE, H. and CARTER, A. (1977) *Excavations in King's Lynn 1963-1970* (Society for Medieval Archaeology Monograph 7) London.

CLASON, A.T. (1978) Voorwepen uit been en gewei. *Spiegel Historiael. Maandsblad voor Geshiedenis en Archeologie*, 13, 294-7.

CLASON, A.T. (1980) Worked bone and antler objects from Dorestad, Hoogstradt 1. In *Excavations at Dorestad 1 The Harbour: Hoogstradt 1* Nederlandse Oudheden 9 Amersfoort, 238-47.

CLUTTON-BROCK, J. and ARMITAGE, P. (1977) Mammal remains from Trench A. In T.R. Blurton, Excavations at Angel Court, Walbrook, 1974. *Transactions of the London and Middlesex Archaeological Society*, 28, 14-100.

CNOTLIWY, E. (1956) Z badan nad rzemisolem, zajmujacym sie obrobka rogu i kosci na Pomorzu Zachodnim we wczesnym sredniowieczu. *Materialy Zachodnio-Pomorskie*, 2, 151-79.

CNOTLIWY, E. (1958) Wczesnosredniowieczne prezdmioty z rogu i kosci z Wolina, ze stanowiska 4. *Materialy Zachodnio-Pomorskie*, 4, 155-240.

COWGILL, J., de NEERGAARD, M. and GRIFFITHS, N. (1987) *Knives and Scabbards* (Medieval Finds from Excavations in London 1) London: HMSO.

DREW, J.H. (1965) The horn comb industry of Kenilworth. *Transactions and Proceedings of the Birmingham Archaeological Society*, 82, 21-7.

EKMAN, J. (1973) *Early Medieval Lund. The Fauna and the Landscape* (Archaeologica Lundensia 5) Lund.

FISHER, F.J. (1936) *A Short History of the Worshipful Company of Horners* London: printed privately.

GOODER, E., WOODFIELD, C. and CHAPLIN, R.E. (1964) The walls of Coventry. *Transactions and Proceedings of the Birmingham Archaeological Society*, 81, 88-138.

HARMAN, M. (1979) The mammalian bones. In J.H. Williams, *St Peter's Street, Northampton: Excavations 1973-1976* Northampton: Northampton Development Corporation, 328-32.

HOLDSWORTH, P. (1976) Saxon Southampton: a new review. *Medieval Archaeology*, 20, 26-61.

KENWARD, H.K. et al. (1978) The environment of Anglo-Scandinavian York. In *Viking Age York and the North* ed. R.A. Hall (Council for British Archaeology Research Report 27) London, 58-70.

KURNATOWSKA, Z. (1977) Horn-working in mediaeval Poland. In La formation et le développement des métiers au moyen âge (Ve°-XIVe° siècles) ed. L. Gerevich. Budapest: Akademiai Kiado, 121-5.

LEPIKSAAR, J. (1975) Über die Tierknochenfunde aus den mittelalterlichen Siedlungen Südschwedens. In Archaeozoological Studies ed. A.T. Clason. Amsterdam, Oxford, New York: North Holland/American Elsevier, 230-39.

MACGREGOR, A. (1978) Industry and commerce in Anglo-Scandinavian York. In Viking Age York and the North ed. R.A. Hall (Council for British Archaeology Research Report 27) London, 37-57.

MACGREGOR, A. (1982) Anglo-Scandinavian finds from Lloyds Bank, Pavement, and other sites. In The Archaeology of York 17, ed. P.V. Addyman. London: CBA, 67-174.

MACGREGOR, A. (1985) Bone, Antler, Ivory and Horn: the Technology of Skeletal Materials since the Roman Period London: Croom Helm.

MACGREGOR, A. and CURREY, J. (1983) Mechanical properties as conditioning factors in the bone and antler industry of the 3rd to the 13th century AD. Journal of Archaeological Science, 10, 71-7.

MANN, J.E. (1982) Early medieval Flaxengate 1: objects of antler, bone, stone, horn, ivory, amber, and jet. In The Archaeology of Lincoln 14, ed. M.J. Jones. London: CBA, 1-68.

MEYER, O. (1979) Archéologie Urbaine à Saint-Denis. Saint-Denis: Maison des Jeunes.

MULLER-USING, D. (1953) Über die frühmittelalterlichen Geweihreste von Wollin. Säugetierkundliche Mitteilungen 1, 64-7.

NODDLE, B.A. (1975) A comparison of the animal bones from 8 medieval sites in southern Britain. In Archaeozoological Studies ed. A.T. Clason. Amsterdam, Oxford and New York: North Holland/American Elsevier.

O'CONNOR, T.P. (1982) Animal bones from Flaxengate. In The Archaeology of Lincoln 18 ed. M.J. Jones. London: CBA, 1-52.

O'CONNOR, T.P. (1984) Selected groups of bones from Skeldergate and Walmgate. In The Archaeology of York 15, ed. P.V. Addyman. London: CBA, 1-60.

O RIORDAIN, B. (1971) Excavations at High Street and Winetavern Street, Dublin. Medieval Archaeology, 15, 73-85.

POLLER, T. (1980) Die Herstellung von dünnen, klaren Hornblättern. Maltechnik-Restauro, 86, 124-5.

PRUMMEL, W. (1978) Animal bones from tannery pits of 's-Hertogenbosch. Berichten van de Rijksdienst voor het Oudheidkundige Bodemonderzoek, 28, 399-422.

PRUMMEL, W. (1982) The archaeozoological study of urban medieval sites in the Netherlands. In *Environmental Archaeology in the Urban Context* ed. A.R. Hall and H.K. Kenward (Council for British Archaeology Research Report 43) London, 117-22.

PRUMMEL, W. (1983) *Excavations at Dorestad 2 Early Medieval Dorestad: an Archaeozoological Study* (Nederlandse Oudheden 11) Amersfoort: ROB.

REICHSTEIN, H. (1969) Untersuchungen von Geweihresten des Rothirsches (Cervus elaphus L.) aus der frümittelalterlichen Siedlung Haithabu (Ausgrabung 1963-1964). In *Berichte über die Ausgrabungen in Haithabu* 2 ed. K. Schietzel. Neumünster 57-71.

ROGERSON, A. and DALLAS, C. (1984) *Excavations in Thetford 1948-59 and 1973-80* (East Anglian Archaeological Report 22): Wachholtz, Gressenhall.

SCHOKNECHT, U. (1977) *Menzlin, ein frühgeschichtlicher Handesplatz an der Peene* (Beiträge zur Ur- und Frühgeschichte der Bezirke Rostock, Schwerin und Neubrandenburg 10). Berlin.

STEWART, I.J. and WATKINS, M.J. (1984) An 11th-century bone *tabula* set from Gloucester. *Medieval Archaeology*, 28, 185-190.

ULBRICHT, I. (1978) *Die Geweihverarbeitung in Haithabu* (Die Ausgrabungen in Haithabu 7). Neumünster: Wachholtz.

ULBRICHT, I. (1984) Die Verarbeitung von Knochen, Geweih und Horn in mittelalterlichen Schleswig. In *Ausgrabungen in Schleswig, Berichte und Studien* 3, ed. V. Vogel. Neumünster: Wachholz.

WALL, S.M. (1981) The animal bones from the excavation of St. Mary of Ospringe. *Archaeologia Cantiana*, 96, 227-66.

WATSON, J. (1984) Organic material associated with metal objects. In C. Hills, K. Penn and R. Ricketts, *The Anglo-Saxon Cemetery at Spong Hill, North Elmhalm Part III: Catalogue of Inhumations* (East Anglian Archaeology Report 21) Gressenhall, 157.

WEIGEL, C. (1698) *Abbildung der ... Haupt-Stände von denen Regenten ... bis auf alle Künstler und Handwerker* Regensburg.

WELCH, C. (1916-23) *History of the Cutlers' Company of London.* London.

WENHAM, L.P. (1964) Hornpot Lane and the horners of York. *Annual Report of the Yorkshire Philosophical Society*, 25-36.

WINKELMANN, W. (1977) Archäologische Zeugnisse zum frühmittelalterlichen Handwerk in Westfalen. *Frühmittelalterliche Studien*, 11, 92-126.

ANIMAL REMAINS AND THE TANNING TRADE

Dale Serjeantson

Treating animal skins to make them more durable has been practised for thousands of years; and tanning appears to have been a specialised skill carried out by full-time craftsmen in the earliest towns. There were tanners in Mesopotamia, Ancient Egypt, and the Roman empire, where leather was such a crucial raw material for the Roman army that regiments had their own tanneries (Waterer 1956; Forbes 1957, 52; Thomson 1982, 144). Here I shall review some of the evidence for tanneries which have been found on excavations, discussing in particular animal bones which may be associated with leather treatment.

There is a range of English terms which have been used by leather workers since the middle ages, which are now no longer widely familiar and it is helpful to define some of them. Skins of large animals such as cattle, horses and camels are 'hides', while those of smaller animals such as sheep, goats, calves and dogs are 'skins' (Waterer 1956). Tanneries today and in the past usually specialised in the treatment of one or other. Hides and skins were treated by the tanner; there was also a form of tanning small skins using alum and salt which is carried out by a tawyer or whittawyer. When the treatment includes the removal of the hair or wool the end product is leather. If the hair or wool is left on the end product is a fur or skin.

The skinner or furrier controlled the trade and treatment of skins of animals whose 'pelts' or 'peltry' were more valuable with the hair or fur on. He bought the pelts, treated them himself or had them treated by a tanner or tawyer, and made them up into garments or trimmings. At some periods an important part has been played by the fellmonger: he is a middleman, who buys hides, skins and pelts from the butchers and farmers and sells them on to the tanneries. In the middle ages pedlars or chapmen acted as middlemen buying the pelts of the small fur bearing animals from foresters, huntsmen and villagers for sale to skinners (Veale 1966), MacGregor (this volume) describes how antlers must have been procured in a similar fashion.

USES OF HIDES AND SKINS

It would be impractible - and indeed irrelevant - to discuss all the products for which leather was used. It is only relevant here where the manufacturers of certain products uses the skins of certain species, or animals of a restricted age range. Among such examples are the use of calf skins for the best vellum, the highest quality being the skin of a stillborn or newborn calf (Diderot & Alembert 1751-8). The Victoria County History for Surrey relates that in the nineteenth century calfskins also had a special use for army chests. Goat and dog skins were used for gloves; suitably treated dog skins were used as fishing floats in Scotland (Shepherd 1979). Parchment was made from goat, sheep or calf (Peignot 1812). Deerskin, goatskin and pigskin, as well as vellum, were used for bookbinding.

SPECIES	NAME	COLOUR	SOURCE
Squirrel (*Sciurus vulgaris*)	Miniver Gris Pople Ruskyn Stranling Rovere Vair	grey ?red black & white	(Russia, Scandinavia (Baltic, including (ultimately Finland
Marten (*Martes* spp.) (*Martes foina*)	Sable Foynes	brown	Northern countries
Polecate or Foumart (*Putorius putorius*)	Fitch Fitchew		British Isles
Beaver (*Castor fiber*)			Wales, Scotland
Fox (*Vulpes vulpes*) ?(*Alopex lagopus*)		black red white	Britain. Also exported. Northern Europe
Cat (*Felis domesticus*)			Britain Also exported
Civet or Genet (*Genetta genetta*)	Genette		Southern Europe
Leopard (*Felis pardus*)			Portugal, ultimately from Guinea
Lamb (*Ovis aries*)	Budge	black, white	Spain, Ireland England (bought from manorial stewards) Also exported
Stoat (*Mustela erminea*)	Ermine Miniver	white white	
Weasel (*Mustela nivalis*)	Lettice	white	
Rabbit (*Oryctolagus cuniculus*)	Coney	black most valuable	England (raised in warrens) Spain. Also exported.
Hare (*Lepus europaeus*)			Britain Also exported
Mouse (*Mus/Apodemus*/etc)	Mouseskin		
Dormouse (*Glis glis*)	Loivre Lemon		
Mole (*Talpa europea*)			Britain

Table 1. Some of the furs used in England in the middle ages. The colours, contemporary names, and place of origin are given where known. Data mainly from Veale, 1966.

FURS

The documentary evidence for the fur trade in the middle ages in England has been examined by Veale (1966). She shows that in medieval Europe furs were worn in complete garments or as trimmings, especially by the rich, and only became less common with the invention of quilted damask and the trend towards keeping houses warmer in the sixteenth century.

Furs have been valued as exchange goods since early times. There is documentary evidence for a fur trade from Scandinavia from the tenth century, but it is likely to have developed there earlier in the Migration Period (Anderson 1981). Fur imports to Britain from long distances are documented from the thirteenth century. In Britain the types of furs imported and worn changed over the centuries, either in response to changes in fashion, as Veale suggests, or to changes in availability. A list of the furs handled, based on information in Veale's study, is shown in table 1. Squirrel fur, especially 'miniver' and 'gris', was most often used in the Court in the fourteenth century, but by the fifteenth century sable was more common. Budge (lambskin) became more common in the sixteenth century; and it is hard to see this as other than a response to the fact that the numbers of people who could afford to wear furs was increasing, and the supply of pelts of wild mammals could not meet with the demand. No doubt the fur bearing species were under pressure: the beaver became extinct in England and Wales in the Middle Ages, and this must at least partly have been due to the demand for beaver furs.

Very large numbers of skins were used in the thirteenth century, many of which were imported. In the Royal Household accounts for 1285-8, 119,300 squirrel skins appear. This immense number is explicable partly if we realise that a 'cote', a short garment, used 366 skins and a robe used thousands. Even strips of squirrel paws, 'poots' or 'potes', were sewn into linings or trimmings.

BONE EVIDENCE FOR THE FUR TRADE

It has been pointed out (Anderson 1981) that "it is in the nature of fur trading to leave few direct and unequivocal traces of its existence in the archaeological record". However, some physical evidence may be found.

The best evidence is traces of cut marks which have resulted from skinning, on the bones of species not usually eaten. Some particularly good examples from a mesolithic site are the remains of pine marten (Martes martes) from Tybrind Vig (Trolle-Lassen 1986). There are skinning cuts on the skull, and the martens were apparently not used for food. In an urban context, some dog skeletons found in a Roman well in Eastbourne, Sussex, have cuts on the nasal bones.

However, the absence of skinning cuts does not mean that the animal was not skinned, since it is possible to skin an animal and leave no trace on the bones. In this case other evidence may be suggestive. Unexpectedly large numbers of juvenile animals may suggest that the species was exploited for its skin. Sixteen out of 25 cats from excavations at Kings Lynn were juvenile or immature, possibly for this reason (Noddle 1977) and the many bones of immature cats in medieval Exeter are thought to have had the same origin (Maltby 1979, 86.)

Figure 1. Sixteenth century illustration of Laplanders wearing fur coats made from pelts on which the tails remain (from Olaus Magnus, Historia de Gentibus Septentrionalis.

Figure 2. A currier's workshop, from Diderot's Encyclopaedia.

The paws and the tail were sometimes left on the fur with the bones still attached. Figure 1 shows Laplanders wearing furs made up from pelts to which the tails are still attached. Where squirrel 'pootes' were used for trimming, the paw bones must have been left in. The origin of the squirrel bones from a cess-pit at Bedern, York (O'Connor 1984) may be the trimming of a fur robe. Foot bones with skins are discussed in more detail below.

If paw bones, but not other parts of the skeleton, are found it is important to consider whether they have come from a pelt or skin; and if they are from a species not native to the area they may be imported. The earliest written references to rabbit skins in England, in 1221, may be to imported skins rather than home produced ones, but the first warrens were set up soon after that date.

Legislation and restrictive practices governed who could trade in and wear furs. The sumptuary laws of Medieval Britain which attempted to prescribe appropriate wear for those of different social status were much concerned with the wearing of furs. Legislation also controlled the guilds and who was permitted to carry out different activities and trades. For example, in Norwich in 1568 butchers had to sell sheepskins to local craftsmen and not to strangers. In the fourteenth century Thomas Legge, a skinner who got rich by supplying the court with furs and became Lord Mayor of London, succeeded in having legislation passed which forbade tanners and glovers to buy skins of lambs killed within the town; they were required to buy them from the skinners (Veale 1966).

THE TANNING PROCESS

Tanning involves several stages, which only varied in detail in different places up until the nineteenth century (Wilson 1941, Forbes 1957, Waterer 1968, Thomson 1982). Thomson comments that a Roman tanner would have found himself quite at home in an eighteenth century tannery.

The processes are:

(1) liming, in which the skins are soaked in a solution containing lime and/or ash, and often urine, which assists dehairing.

(2) scraping or scudding, in which the hair, the epidermis and any remaining subcutaneous flesh and fat are removed from the skin.

(3) either (a) bating or puering, in which skins are soaked in an alkaline solution containing trypsin, a pancreatic enzyme which softens the skin and makes it receptive to the action of tanning agents (Wilson 1941). Formerly substances used for this included dog excrement, which gives a weak solution, or pigeon dung, which gives a stronger one.

(b) or drenching, in an acid solution closer to fermentation, which used such materials as stale beer or urine;

(4) tanning, for which there were formerly two principal processes. Tanning with vegetable tanning agents produces the most durable results; but treatment with alum and salt, 'tawing', was also in use from the middle ages. Mineral tanning agents, which are used almost exclusively today, have been common only since the nineteenth century.

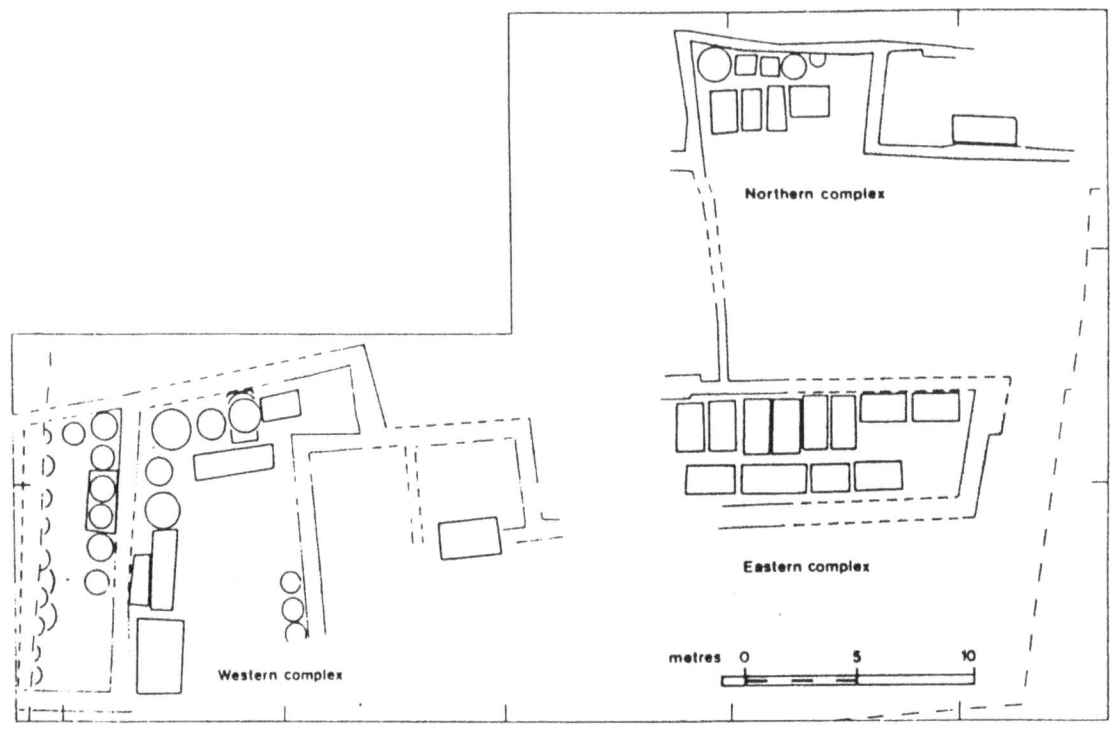

Figure 3. Plan of tanning pits at the Green, Northampton. The northern and western complexes contain both round and rectangular pits. (Reproduced with permission from Shaw 1984).

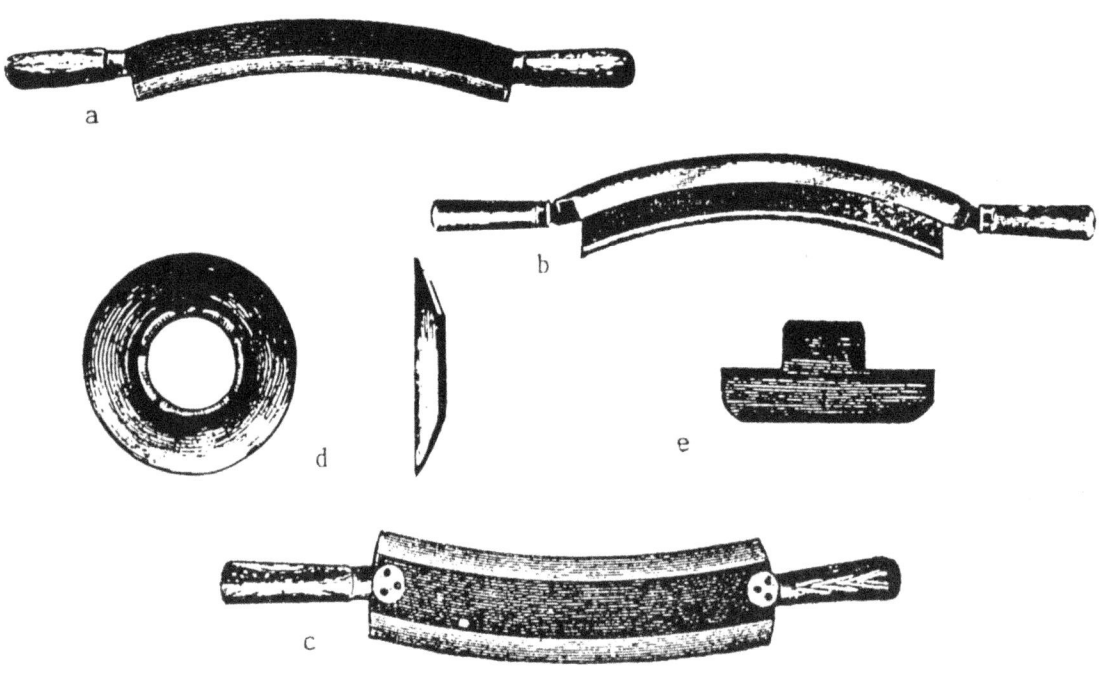

Figure 4. Eighteenth and nineteenth century tanners and currier's tools. (a) unhairing knife (b) scudding knife (c) fleshing knife (d) lunette (e) currier's knife. (a), (b) and (c) from Bennett 1919; (c) and (d) from Diderot's Encyclopaedia.

(5) After the hides and skins have been treated they are dressed, 'fatliquored' or 'curried' with oils to make them supple. This may be done in a number of ways, and was often done by specialist curriers. Figure 2 shows the activities of the currier in the eighteenth century as illustrated in Diderot's Encyclopaedia. Men are scraping imperfections off the skins over flat tables or over an angled beam, beating a skin, and trampling skins to make them supple (Waterer 1956).

Alternatives to tanning are treatment with oils, a process known as 'chamoising', after which the leather obtained was supple but not very durable, and parchment making, where the skin is limed, scraped and dried, but not treated further (Gaur 1984). Parchment too is less resistant to decay than leather.

The leather was then made up into goods by the various specialist craftsmen, such as the cordwainers, shoemakers and the saddlers. These craftsmen formed themselves into powerful guilds in medieval Europe, and there is an extensive literature about their craft, in contrast with the early stages of the process, which is ill-documented.

PHYSICAL EVIDENCE FOR TANNERIES

As tanneries need a supply of water they were often sited near a river or stream (Keene 1982). "Without great store of running water they cannot dress the same" (Landsdowne Ms 74 f 52 quoted by Salzman 1913, 173). Written evidence for the presence of tanneries can be found in records when they were fined for discharging their foul 'ooze' or 'wooze' on the highway.

Former tanneries will most often be identified from the presence of tanning pits. Pits for skins are about 4' (1.2m) in length and those for hides are larger, up to 6' (1.8m) and more (Douglas 1956). They may be rectangular or round and lined with stone, clay, wood, or a combination of these materials. Many have now been recognised in urban excavations. Pits of the Roman period, together with debris, were found at Augst (Augusta Rarica) in Switzerland (Schmid 1969). Post-medieval examples include adjacent sites with at least 55 pits at St Peter's Street and Walmgate, York (Brinklow 1984). Both circular and rectangular pits were found at the Green (Figure 3).

At the Green, Northampton, some evidence for stave and plank linings to the pits was found, and deposits of lime and also ash, probably used in the lining process (Shaw 1984). The tanning pits recently excavated at the brooks, Winchester (Times 29.10.1987) were near a building which is thought to be a dove-cot; and it is tempting to regard the proximity of tanning pits and a source of material for bating the skins as more than fortuitous.

TOOLS

The tools of the tanner may survive, particularly those made of metal. The unhairing knife, scudding knife and fleshing knife illustrated in Figure 4 have a characteristic curved blade which ensures that if the knife slips it will not accidentally nick the skin and damage it (Thomson 1982). A wide range of the tools are illustrated in a recent compendium (Salaman 1986). The curved shapes have an ancestry very much more ancient than medieval

Europe, and will be familiar to any student of flint tools.

OTHER TRACES

Some materials will only survive in waterlogged or other anaerobic conditions.

Scraps of leather offcuts have survived at many sites, including Coppergate, York, Freienstrasse, Basel, Switzerland (Schmid 1973) and sites in Roman London. A few of the scraps from London have inscriptions stamped on the leather (Rhodes 1987).

Some waterlogged plant material have been found which have been interpreted as tanning materials. Very many plants can be used as tanning agents, as any herbal will show, though oak bark was most common in temperate Europe as it is higher in tannins that other native species (Reed 1966). It has been found at the two Swiss sites already referred to, Augst and Freienstrasse.

Certain insects may be revealing; insect remains survive not only in waterlogged conditions but also as a result of replacement of the soft parts by calcification. The insect fauna from a 16th or 17th century waterlogged pit in Southwark probably indicated tanning in the vicinity: the most numerous beetle was Trox scaber, which feeds on hides and dry carcasses, and the only woodland beetles, Siagonum quadricorne Kirby & Spence and Abraeus granulum (Er.), are two species which live under the bark of deciduous trees, and were thought to have been introduced with bark for tanning (Girling 1979).

EVIDENCE FROM ANIMAL BONES

Some animal bones may be associated with skins and hides and their presence in an archaeological context may therefore indicate the presence of a tannery or a connection with the leather trade. The question hinges on whether bones were left attached to the skin when the animal was slaughtered and skinned, and, if so, which bones?

It appears that in the past either the horn cores, or the feet, or both, were left on the skin. Today no bones remain on the skins delivered to tanneries: this was affirmed by several Northamptonshire tanneries which were contacted. Text books for butchers (e.g. Gerard 1949, Aten et al. 1955) make no reference to the practice. It has been pointed out (MacGregor 1978) that the weight of the skull or part of it and/or the feet would add considerably to an already heavy hide or skin. If therefore they were attached to the skin in the past there would have to have been advantages for the tanner.

Both horns and foot bones have been found associated with tanneries. One very clear example of foot bones at a tannery site is Walmgate, York (O'Connor 1984), where a series of eighteenth century tanning or tawing pits was excavated. Many hundreds of metapodials and phalanges of sheep – over 85% of the bones – were found in pits or scoops at the site. Numbers of phalanges were higher than metapodials, so, as O'Connor has argued, it appears that some skins were delivered with both metapodials and phalanges attached and others with phalanges alone. The asssociation was clear, too, at the excavations discussed by Schmid (1969, 1972, 1973). At St. Peters Street, Northampton, the excavated series of pits replaced a single earlier pit (Williams 1979). The fill of the earlier pit contained 45 metapodials and 136 phalanges of sheep (Harman 1919). They were interpreted as butchery waste; and here I suggest an alternative interpretation for their origin.

Horn cores have been found at several known and putative tanneries, including the Green, 's' Hertogenbosch, Netherlands (Prumel 1978), Freienstrasse and Augst. The association of horn cores with tanneries is also discussed by MacGregor (this volume).

There is some documentary and anecdotal evidence that the feet, the horns and also the tail were sometimes left on the skin. A German woodcut of a tanner shows hides hanging in the workshop with skull and tail attached (Figure 5). Prummel (1978) cites a Dutch manual for leatherworkers which states that "the first task is to remove the horns and tails".

The Managing Director of a tannery in Colyton, Devon, which still uses oak bark and traditional methods, Mr. D.F. Baker, wrote: "The writer remembers when there were horns on the hides when bought and also foot bones, but this must have been 50 or 60 years ago at least. Always in those days before tanning, the horns and foot bones were removed from the hide and they were collected from the tanyard as and when a load was available". The Conservator of the Leather Conservation Centre, Northampton, confirmed that "until the middle of the 19th century hides were bought from the butcher with the hooves, horns and other appendages attached". A nineteenth century watercolour of a skin-cart from Kingston (Figure 6) shows the skins with what appear to be feet still attached (Freelove 1979). At least one village butcher and slaughterman in Kent leaves the feet attached to the skins for the fellmonger (J. Wakeford, pers. comm.). Navajo Indians have also said that the fellmonger liked to get the sheepskins with the phalanges still attached (Binford & Bertram 1977, 92, 94).

Disproportionately high numbers of foot bones are quite often found on urban sites. An eighteenth century pit on the Knapp Drewett site in Kingston was filled almost exclusively with calf foot bones (Serjeantson et al. 1987). These are of interest for two reasons. Firstly, the numbers are so great - the pit probably contained the bones of about 60 animals - that some association with crafts and trades seems a more likely explanation for their presence than slaughterer's waste. Secondly, the calves from which the bones came had all been slaughtered at much the same age of about four months. There are cuts on the surface of the metatarsal and metacarpal which show that the leg was usually dismembered at that point. The age at death suggests that the calves were sold at the optimum age which combined the provision of cows' milk for the town and veal for consumption. It is argued that the presence of so many bones of the same types makes them unlikely to be butchery waste; and more likely to be connected with the tanning trade. There are known records of a tannery at the site. The bones may have been from skins assembled by a fellmonger.

Numerous sheep phalanges were found among the animal bones from a pit at Bewell House, Hereford (Noddle 1985). Metapodials have also been found in high numbers at some sites. In Figure 7 the proportions of the main bones in the skeleton from three assemblages are shown in a histogram, and they are compared with the number of bones expected, based on a recent ethnographic sample from a Hottentot village (Brain 1976). The Hottentot

Figure 5. German woodcut of 1568 of a tanner at work, showing a cow hide hanging up with the skull and tail still attached.

Figure 6. Nineteenth century skin cart from Kingston Upon Thames, showing skins with feet attached. (Redrawn from Freelove 1979).

bones show the relative numbers in which the bones survive, if the whole carcass was present and consumed on the site; and the histogram shows that metapodials are disproportionately high at the three sites. They are: (1) a small sixteenth century assemblage from 29 Thames Street, Kingston-upon-Thames, (Serjeantson n.d.), (2) the well at Rudston Roman Villa (Chaplin & Barnetson 1980) and (3) eighteenth century deposits in Rye town ditch (Kyllo 1981).

At Rye there were offcuts of leather with the bones from the ditch (Hadfield 1981), confirming that the rubbish dumped in that spot was associated with leather working. Thames Street had several tanners as tenants of the houses in the fifteenth century, and a century later No. 29 belonged to the first recorded master tanner in Kingston, one Obadiah Wicks. It is near the site of the town tanneries which survived until the 1940s.

This review of associations between horn cores, foot bones and tanneries has shown that they may indeed be found at tannery sites. But **must** a collection of this nature be tannery waste? The horns have no meat, and the foot bones have little, so are discarded early in the butchery process. Indeed a high proportion of skull fragments and foot bones may be diagnostic of waste from primary butchery, as contrasted with waste from food consumption (cf Levitan, this volume). In small quantities this undoubtedly is so. However, when the disproportion in the sample is very great we should also consider as an alternative that the origin lies in a craft process, in this case the leather trade. As Chaplin (1971) wrote: "In a slaughteryard skulls and feet would only be found if they are not being sold". Crafts and trades were highly organised in towns from early times and in such a context no potential raw material was wasted.

WHY?

The reason why horns, feet and the tail were left attached to the skins is not entirely clear, and more than one reason has been put forward. After observing the association between horns and foot bones and tannery sites, Schmid investigated the question. One winter in Graubunden, Switzerland, she saw "two goat skins drying in the winter sun. They were hanging inside out lengthwise over a pole. The feet, in which the end of the metapodial bones and the toes with the hooves were still present, were tied together and between them there hung a board, the weight of which provided the necessary tension to prevent the skin from shrinking. The head was not cut off completely, but part of the skin of the head was present with the long twisted horns attached". The farmer said "He had learned from his father how to skin a goat this way". The reason given for leaving horns on the skin was so that "the tanner can easily know the age of the animal" (Schmid 1974, 10).

A more powerful argument for leaving these bones attached must lie in the fact that all have their uses. Deposits of caudal vertebrae have not been recorded so I will not speculate about tails, but horn was such a valued commodity in its own right (MacGregor 1978 and this volume) that it is hard to imagine that horns were ever discarded before the horny sheath was removed for use. When the tanner received skins with the horns still attached he surely passed the horns on to the horn worker. Both usually worked near each other in the part of the town devoted to the smellier craft activities.

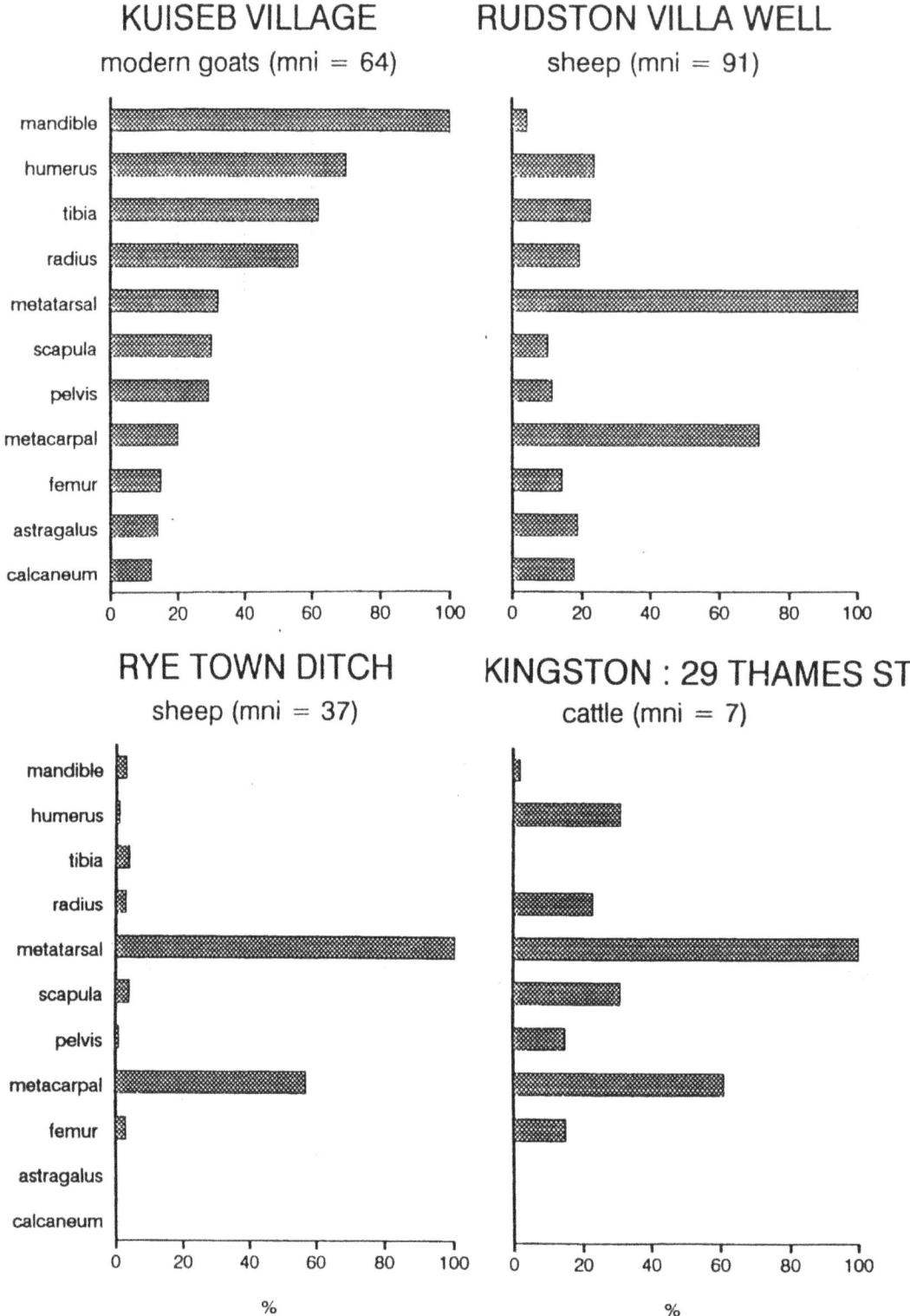

Figure 7. Histogram showing representation of the main bones at Rudston Villa, 29 Thames Street, Kingston-Upon-Thames and the town ditch, Rye. The bone proportions are compared with those from a recent ethnographic sample which shows the relative numbers expected when the complete carcass was consumed on site. The three archaeological sites have a disproportionately high number of metapodials. The Kuiseb example is adapted from Brain 1976, 114.

I would like to propose here a reason why the tanner wanted the foot bones. As mentioned above, 'chamoising' or oil treatment of skins requires fine oils, as does currying leather after tanning. The thinnest animal fats or oils, those highest in oleic acid, are obtained from bones from the lowest part of the leg (Binford 1978, table 1.11). Neatsfoot oil is the finest and thinnest animal oil obtainable, with the exception of oils from some marine mammals (Encyclopaedia Britannica 1903 744, 747). It has a lower melting point than other animal fats, and remains liquid below 0°C. In its purest form it is made only from the phalanges of cattle, but oil from the feet of horses, sheep, and goats is also sometimes sold as neatsfoot oil, and in America in the nineteenth century the metapodials as well as the phalanges were used (Thorpe 1922). It is produced today in back yard workshops, where batches of cattle feet bought from a knacker are boiled up in malodorous vats (M. Jeffries, pers. comm.) and the oil then bought by wholesalers and retailed at a high price as a leather dressing, especially for horse gear. Its earlier use as a fine oil for delicate machinery has now been superceded by mineral oils.

As the main use of neatsfoot oil in early times was for leather dressing it seems likely then that the tanner wanted the feet to ensure a supply of the oil either for himself or for the currier. This is corroborated from an American source (Binford & Bertram 1977). The Navajo gave as a reason why the fellmongers wanted sheep skins with the feet on: "the grease of the foot" is used by the tanners.

Some at least of the calves' feet from Kingston were articulated when they were discarded, and in discussing that assemblage (Serjeantson et al 1987) it was suggested that the bones would not remain articulated if they had been rendered for oil. However a recent description of the production of cow heel and neatsfoot oil calls this into question. The daughter of a butcher working in Coventry in the 1930s described to a **Guardian** reporter how her father obtained and prepared tripe, cows' heel and neatsfoot oil. He "would insert a knife at the crucial spot to extract the neatsfoot oil", which he then sold in his shop alongside the tripe and cows' heel (Figure 8). It is therefore possible to obtain the oil without disarticulating the foot bones.

CONCLUSIONS

In this paper I have surveyed the physical remains which may be found at tanning sites and argued that groups of foot bones should be considered as one part of the evidence. However, the nature of rubbish disposal is such that the waste products from leather treatment, such as bones and leather offcuts, will not necessarily be disposed of in the primary context, that is at the site where treatment took place. Though some groups of foot bones **have** been found in unequivocal association with tanneries, others have not. It is likely that Rye town ditch, for instance, contained bones and leather offcuts because it was a site chosen for secondary disposal of waste from the town. In addition, there were undoubtedly circumstances in which butchers sold the skins and hides without the feet attached and disposed of the bones of the feet with their other waste. Nevertheless, the presence of foot bones, as well as the other physical remains from tanning, may alert the excavator to tanning activities in the vicinity.

Both records and excavations have shown that in small towns the trades which made use of animal products, especially the butchers, tanners and

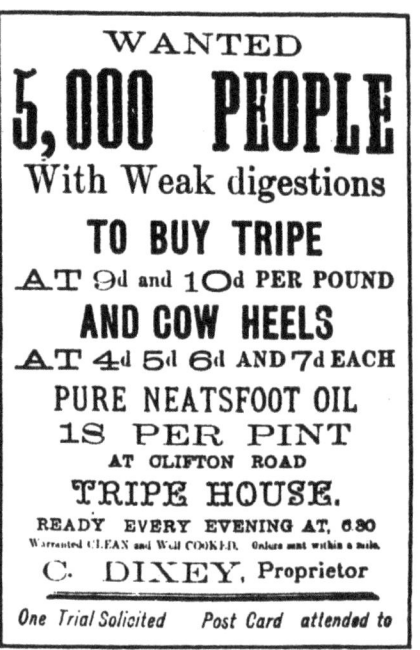

Figure 8. Poster of the 1930s advertising tripe and neatsfoot oil for sale in the same shop.

horners, were found together, and it is not easy to distinguish their rubbish. The waste products from crafts associated with animals are often smelly and unpleasant. They are rarely left at the primary location, but were buried possibly at a distance from where they were used and worked. As other contributors to this book have stressed, distinguishing the waste products of craft activities in towns is a valuable means by which archaeologists can gain insight into town life as revealed by excavation.

ACKNOWLEDGEMENTS

I would like to express my thanks to the following people who have provided helpful information: Mr J. Baker; Mark Bracegirdle, Mr M. Jeffries of MCP; Tony Legge, Terry O'Connor, R. Thomson, Dr. E. Schmid. Miss Joan Wakeford gave information about the early history of Kingston upon Thames and made many useful comments. I am grateful to Pat Stevens for allowing me to examine the dog skulls from Eastbourne and for permission to refer to them.

REFERENCES

ANDERSON, A. (1981) Economic change and the prehistoric fur trade in Northern Sweden; the relevance of a Canadian model. Norwegian Archaeol. Rev. 14 (1), 1-38.

ATEN. A., INNES, R., FARADAY and KNEW, E. (1955) Flaying and Curing of Hides and Skins as a Rural Industry. Rome: F.A.O. Agricultural Development Paper 49.

BENNETT, H.G. (1919) The Manufacture of Leather. London: Constable.

BINFORD, L.R. (1978) Nunamiut Ethnoarchaeology New York: Aacdemic Press.

BINFORD, L.R. & BERTRAM J. (1977) Bone frequencies and attritional processes. In For Theory Building in Archaeology, ed. L.R. Binford. New York: Academic Press, 77-153.

BRAIN C.K. (1976) Some principles in the interpretation of bone accumulations. In eds G. Ll. Isaac & E.R. McCown, Human Origins New York: W.A. Benjamin, 97-116.

BRINKLOW, D.A. (1984) Walmgate: the archaeology. In Selected Groups of Bones from Skeldergate and Walmgate by T.P. O'Connor. The Archaeology of York 15.1 London: Council for British Archaeology, 30-31.

CHAPLIN, R. (1971) The Study of Animal Bones from Archaeological Sites. London: Academic Press.

CHAPLIN, R. & BARNETSON, L.P. (1980) The Animal Bones. In Rudston Roman Villa by I.M. Stead. London.

DOUGLAS, G.W. (1956) Survey of the Production of Hides, Skins, and Rough Tanned Leathers in India, Pakistan, Ceylon and Africa London: British Leather Manufacturers Research Association.

DIDEROT, D. & ALEMBERT, J.d' (1751-8) Encyclopédie ou Dictionnaire Raisonné des Sciences, vol. 6. Paris

ENCYCLOPAEDIA BRITTANICA (1903) Tenth edition. Volume 6.

FORBES, R.J. (1957) Studies in Ancient Technology V. Leiden: E.J. Brill.

FREELOVE, W.F. (1979) Victorian Horses and Carriages ed. D.L. Jens Smith. Kingston.

GAUR, A. (1984) Exhibition notes for A History of Writing London: British Library.

GERARD, F. (1949) The Book of the Meat Trade Vol. 1. London: Caxton Publishing Company.

GIRLING, M. (1979) The entomological evidence for tanning from a post-medieval pit at Southwark. Unpublished report to the Ancient Monuments Laboratory, No. 2735.

HADFIELD, J. (1981) An excavation at 1-3 Tower Street, Rye, East Sussex. Sussex Archaeol. Collections 119, 222-225.

HARMAN, M. (1979) The mammalian bones. In St Peter's Street, Northampton: Excavations 1973-1976 by J.H. Williams. Archaeological Monograph 2. Northampton: Northampton Development Corporation, 328-32.

KEENE, D.J. (1982) Rubbish in medieval towns. In Environmental Archaeology in the Urban Context, ed. A. Hall & A. Kenward. Council for British Archaeology Research Report 43. London: Council for British Archaeology.

KYLLO, M.J. (1981) The animal bones. In J. Hadfield, An excavation at 1-3 Tower Street, Rye, East Sussex Sussex Archaeol. Collections 119.

MacGREGOR, A. (1978) Bone, Antler, Ivory and Horn London: Croom Helm.

MAGNUS, Olaus (1567) Historia de Gentibus Septentrionalis.

MALTBY, J.M. (1979) The Animal Bones From Exeter 1971-75 Exeter Archaeological Reports 2. Sheffield. University of Sheffield.

NODDLE, B.A. (1977) Mammal bone. In Excavations in Kings Lynn 1963-1970 by H. Clarke & A. Carter. Monograph 7. London: Society for Medieval Archaeology, 378-408.

NODDLE, B.A. (1985) Bewell House: the animal bones. In R. Shoesmith, Hereford City Excavations, Vol. 3. C.B.A. Res. Rep. 56. London: Council for British Archaeology, fiche M8.F1-G14.

O'CONNOR, T.P. (1984) Selected Groups of Bones from Skeldergate and Walmgate The Archaeology of York 15.1 London: Council for British Archaeology.

PEIGNOT, G. (1812) *Essai sur l'histoire du parchemin et du vélin* Paris.

PRUMMEL, W. (1978) Animal bones from tannery pits at s'Hertogenbosch. *Berichten van de Rijkdienst voor het Oudheidkundig Bodermonderzoek* 28 (1978), 399-422.

REED, R. (1966) *Science for Students of Leather Technology* London.

RHODES, M. (1987) Inscriptions on leather waste from Roman London. *Britannia* XVIII, 173-181.

SALAMAN, R.A. (1986) *Dictionary of Leatherworking Tools* c1700-1950. London: George Allen & Unwin.

SALZMAN, L.F. (1913) *English Industries in the Middle Ages.* London.

SCHMID, E. (1969) Knochenfunde als archäologische Quellen. In *Archäologie und Biologie, Forschungsberichte* 15. Deutsche Forschungsgemeinschaft 1968. Wiesbaden: Steiner 100-111.

SCHMID, E. (1972) *Atlas of Animal Bones* Amsterdam: Elsevier.

SCHMID, E. (1973) Ziegenhörner als Gerberei-Abfall. *Korrespondenzblatt der schweizer Gesellschaft für Volkskunde* 5/6, 65-70.

SCHMID, E. (1974) Als Das Gerben noch ein langweriges Geschaft war *CIBA-Geigy-Zeitschrift* 1/74, 8-11.

SERJEANTSON, D., T. WALDRON & S. MacCRACKEN (1987) Veal and calfskin in eighteenth century Kingston. *London Archaeologist* 5, 9. 227-232.

SERJEANTSON, D. (n.d.) *The animal bones from 31 Thames St. Kingston Upon Thames.* Unpublished report. Kingston Museum.

SHAW, M. (1984) Northampton: excavating a 16th century tannery. *Current Archaeology* 91, 241-244.

SHAW, M. (1987) Early Post-Medieval Tanning in Northampton, England. *Archaeology* 40.2, 43-47.

SHAW, M. (forthcoming) *The Excavation of an Early Post-Medieval Tannery and Earlier Remains at the Green, Northampton.*

SHEPHERD, H.D. (1979) Dog bowies: the use of dogskins for fishing floats. *Scottish Studies* 23, 83-86.

THOMSON, R.S. (1982) Tanning: Man's first manufacturing process? *Transactions of the Newcomen Society* 53, 139-155.

THORPE, W. (1922) *Dictionary of Applied Chemistry.* London.

TROLLE-LASSEN, T. (1986) Human Exploitation of the Pine Marten (*Martes martes* L.) at Tybrind Vig STRIAE 24, 119-122.

VEALE, E.M. (1966) *The English Fur Trade in the Later Middle Ages* Oxford: Clarendon Press.

WATERER, J.W. (1956) Leather. In *History of Technology*, ed. Singer & Holmyard. Oxford: Clarendon Press.

WATERER, J.W. (1968) *Leather Craftsmanship* New York: G. Bell & Sons Ltd.

WILLIAMS, J.H. (1979) *St. Peter's Street, Northampton: Excavations 1973-1976* Archaeological Monograph 2. Northampton: Northampton Development Corporation.

WILSON, J.A. (1941) *Modern Practice in Leather Manufacture* New York: Reinhold.

THE USE OF ANIMAL BONES AS BUILDING MATERIAL IN POST-MEDIEVAL BRITAIN

Philip L. Armitage

INTRODUCTION

In 1973, during demolition of an 18th century cottage in Ware, Hertfordshire, members of the local archaeological society (Ware Rescue Group) uncovered a flint-built wall beneath the building, a section of which had suffered structural damage and been repaired using cattle metapodial bones set in mortar as replacements for the dislodged flints (figures 1 and 2). Writing about this discovery, the director of the excavations commented that although he knew of no parallel for the "bone wall" at Ware he was "reminded of a number of cases of horn cores being used for floor bases and drains which have been part of the oral tradition of archaeology but which don't seem to have appeared in print" (Crossby, 1974: 175). Crossby's references to "horn cores" and "drains" were to prove prophetic: in 1978 excavations conducted by members of the Enfield Archaeological Society in north London discovered examples of late 17th/early 18th century agricultural land drains lined with cattle horn cores laid end to end (see Armitage, Coxshall & Ivens, 1980). Two horn cores used in these drains are shown in Figure 3. Enquiries made among other local archaeological groups following the north London find brought to light an earlier (1968) - but hitherto unpublished - discovery of 17th/early 18th century horn core lined land drains in Old Hatfield, Hertfordshire (Harris, 1983, pers. comm.). Also, in 1978, archaeologists from the Museum of London's Department of Urban Archaeology made the remarkable discovery beneath the Cutler Street warehouse complex in the east end of London of two industrial pits (exact function not known) whose sides had been reinforced with cattle horn cores laid in neat rows (Figures 4 and 5), see Armitage, 1982: 102.

Prompted by the archaeological discoveries at Ware, north London and in the east end of London, the author decided to investigate further the archaeological, documentary and surviving architectural evidence for the use of animal bone as building material. A brief synopsis of the preliminary results of this research was presented at the Fourth International Archaeozoological Conference held at the Institute of Archaeology, University of London, April 18th - 23rd 1982.

This paper is a revised and updated version of that synopsis and incorporates information supplied by colleagues during and after the ICAZ meeting as well as documenting more recent archaeological discoveries (up to 1985). A reference list of the examples discussed in this paper is given in the Gazetteer at the end of this volume. A number of examples quoted in the Gazetteer are unprovenanced and/or undated; but they have been included so as to provide the most comprehensive listing possible for the use of field archaeologists and others who may encounter further examples.

Before examining the evidence it is important to first make the distinction between the "rational" and the superstitious/"magical", ie. "non-rational" (as defined by Merrifield, 1969: 103) use of animal bones in buildings and other man-made features.

Figure 1. Eighteenth century cottage, Ware, Hertfordshire: view along the top of the basement wall showing the section which had suffered damage and been repaired with cattle bones set in mortar. Photo: Fred Crosby.

Figure 2. Part of the same wall close up, showing metapodials laid lengthwise across the width of the wall. Photo: Fred Crosby.

"RATIONAL" VERSUS THE "NON-RATIONAL" USE OF ANIMAL BONES

Although this paper is principally concerned with examples of the purely practical purposes to which animal bones have been put in building constructional projects in the past, it is also important that the field archaeologist is able to identify circumstances where the bones sometimes found in ancient or historic man-made structures were originally placed there for some "magical" or superstitious purpose. Therefore the following observations are offered in order to clarify the distinction.

It is well known that animals have had a very long association with the craft of building; and from very early times have often unwittingly played a role in the rituals attending the laying of foundations to houses, bridges and fortifications. The custom of animal sacrifice on the foundations of a new building sometimes persisted in supposedly Christian contexts; according to Swedish tradition, a freshly killed lamb was buried under the altar of the early Christian churches in order to give them "security" (Burdick, 1901: 57). The belief that the bones of animals gave "strength and stability" to the structure of a building continued throughout the Middle Ages and even well into the 18th century in Britain, as evidenced by the quantity of cattle, sheep and horse bones placed beneath the foundations to the piers of Old Blackfriars Bridge in London which was constructed between 1760-69 (see Speth, 1893: 10).

Evans (1966: 55) attempted to explain the reason behind the survival into the medieval and post-medieval periods of the other superstitious practices; reasoning that the followers of these apparently pagan rituals may not necessarily have believed whole-heartedly in them ".... but if good fortune involved merely... throwing a horse-bone into the foundation or secreting an ox-bone in the structure it was better to be on the side of luck. When this had been done, whatever eventually happened, the house would at least _feel_ a more secure place to live in".

As discussed by Merrifield (1969: 102) it is important to recognise that in the Tudor and Stuart periods in Britain "there was a change in the purpose of [continuing the older] superstitious practices. The original idea had been positive - to bring good fortune to the building; the emphasis now was probably the warding-off of evil, which might enter by the chimney or threshold". Merrifield goes on to say that it is against the background of the obsession with counter-witchcraft that one must view "the revival [at that time] of superstition in its most repulsive forms" - as exemplified by the placing of two strangled and two live chickens in a bricked-up recess near the first floor fire-place in Lauderdale House, Highgate Hill, London, sometime during the late 16th or early 17th century (ibid: 101-102). Porter (1969: 180-81, and 1982 _pers. comm._) relates an example of the survival into modern times of one of these counter-witchcraft practices: as late as 1897, in the fenlands of Cambridgeshire, a local builder was observed placing a horse's head (brought from a knacker's yard) in the foundation trench to the Methodist Chapel then under contruction at Black Hourse Drove, a district of Littleport; the head was 'anointed' with beer before being covered with bricks and mortar. When asked the reason for his actions, the builder declared that this was "..an old heathen custom to drive evil and witchcraft away".

It is interesting to speculate what an archaeologist in the future would make of this particular horse's head burial - assuming of course that he or she did not have the benefit of the eyewitness account of the ritual

Figure 3. Horn cores from late seventeenth/early eighteenth century agricultural land drains at Upsell Avenue, Enfield. Top: unimproved longhorn bull. Bottom: medium horned cow. Photo: Trevor Hurst, Museum of London.

as recorded in Porter (1969). Perhaps they would dismiss any notion that this was a foundation rite and opt instead for a much more rational explanation - that the horse skull was intended to serve as a sounding box to improve the acoustic qualities of the church.

The problem of determining the significance of horse skulls (either singly or in large groups) found under the floors in old houses, barns and churches has long been the subject of fierce controversy among folk-lore researchers. Suilleabhain (1945) found from a postal survey conducted in Ireland that the practice of burying horse skulls within buildings was at one time widespread throughout the country. Although most of his correspondents were of the firm opinion that in every case the buried skull (or skulls) served as acoustic resonance vessels that enhanced the sound produced by dancing feet, Suilleabhain argued that this was probably not the original motive behind such burials. He thought it much more likely that the skulls were part of a "foundation sacrifice"; but that this original purpose was probably forgotten "with the passing of the years, and the echo motif supplanted it" (ibid: 50). Sandklef (1949: 43) questioned this assumption made by Suilleabhain, and produced evidence from his own survey of old Swedish customs to show that horse skulls placed beneath the floors of threshing barns were there to give resonance to the floor and so help the threshers in performing their work by enabling them to keep up a regular "singing rhythm" whilst swinging their flails. Nowhere could Sandklef find evidence that the skulls were meant as talismen to ward off evil or bring good luck to the grain harvest; the measures adopted for protecting the contents of the barns from evil forces were on more open display and always took the form of pentagrams and other mystical symbols carved above the doors or on the wall-posts.

Despite the reluctance of certain archaeologists to accept the survival into recent times of the older superstitious rituals (see Merrifield, 1969: 103) there can be no doubt that <u>isolated</u> skulls found in a number of 17th and 18th century houses were indeed meant to protect the building and its inhabitants; the depth of these skulls would have made them totally ineffective as acoustic vessels (see Buchanan, 1956: 60; also Gailey, 1971). However, Sandklef's explanation is equally valid where large numbers of horse skulls have been discovered together immediately beneath and in close contact with wooden floorboards in 17th and 18th century houses and inns; in Britain (see Gazetteer, Section 6.1) there can indeed be very little doubt that under the circumstances the skulls were intended to improve the sound effects in the room; and no "mystical" function need therefore to be invoked to explain their presence.

The discovery of animal bones other than skulls in the walls of 17th and 18th century houses presents the same problem of interpretation; but as with the horse skulls the positioning and quantities of the bones will provide clues as to the original purpose; whether ritual or entirely functional: if the motive was either to bring good fortune or to ward off evil forces, only one or two animal bones should be present. Conversely, the presence of a large quantity of bone would indicate their use in a functional way (eg. as packing or hardcore, see Gazetteer, Section 3.1, also 3.2).

Figure 4. Cutler's Gardens, City of London: post-medieval industrial pit with sides lined with cattle horn cores. The surviving side shows the mode of construction, with horn cores laid in courses bonded by layers of clay. Photo: Trevor Hurst.

Figure 5. Same as 4, view on to the exposed surface of the top of one of the pits. The picture shows clearly the arrangement of the cores, with the tips pointing outwards from the centre of the pit. Photo: Jon Bailey, Museum of London.

ANIMAL BONES AS BUILDING MATERIAL

The survey conducted into the archaeological, documentary and surviving architectural evidence produced a large number and variety of examples of the use of animal bones as building material. These examples are documented in the Gazetteer.

Four important points emerged from the survey:-

1) The practice in Britain of using animal bone as building material is of no great antiquity. None of the dated examples discovered so far is earlier than the 17th century.

2) A preponderance of the examples were located in, or close to large urban centres in the Midlands and south eastern England.

3) In every case, the bones used were the by-products of butchery, tanning or horn-working ie. were waste materials that would otherwise have been carted away and buried in rubbish pits.

4) The practice was mostly confined to a relatively short period, and ceases some time during the late 18th century.

From these points it is possible to place the custom of building in animal bones in its proper historical context; and suggest reasons for its origins as well as speculate on why it was such a short lived phenomenon.

The time period, between the 17th and late 18th century, and the concentration of the examples in the more heavily populated Midlands, south and south eastern districts of England are factors of particular significance: it was from the late 16th century onwards that the cities and towns in this part of Britain experienced an unprecedented growth in their populations. Nowhere was this increase in population more evident than in London, where within the space of only 55 years, the population increased by over 400%, from an estimated 50,000 in 1550 to 225,000 by 1605 (Schofield & Dyson, 1980: 67) and by the second half of the 17th century, the figure was approaching 500,000 (Guildhall Library ref. sheet C40.1/T). Although other urban centres in the same region did not show such a dramatic growth as London they nevertheless experienced significant increases in their populations.

In order to supply the enormous quantities of meat required to feed the populations of these greatly enlarged urban centres, cattle born and raised in the remoter regions of Britain (Scotland, the Lancashire Plain and Wales) were driven in vast numbers 'on the hoof' to the Midlands, and the south and south eastern districts of England. Sheep also were driven over long distances to supply these same urban markets. The slaughter of these animals would have resulted in the production of large quantities of metapodial bones and cattle horn cores (the principal bone elements used as building material). Unfortunately there are insufficient data to calculate the quantities of these bone elements derived from slaughteryards in every city and town where bones have been used as building material. In London, however, some idea of the quantities available for this purpose may be obtained from the following rough estimates of bone waste produced in one year during the period under consideration:-

According to the figures published in M'Culloch's Dictionary of Commerce (quoted in Youatt, 1834: 256) 607,330 sheep were purchased by butchers at Smithfield market in 1737. The slaughter of those animals would have yielded 2,429,320 metapodial bones (this figure included bones from lambs and immature animals which would not have been suitable for use as a building material). Similar calculations may be made for the number of metapodial bones and horn cores produced by the slaughter of cattle in the City of London in the late 17th and early 18th century (Table 1).

The estimated values provide some idea of the size of the problem faced by the butchers, tanners and horn-workers in London, who would have had to arrange for the disposal of this material. Although a proportion of the sheep and cattle metapodial bones would have gone to bone-workers, there would still have been considerable quantities of these bones, and the cattle horn cores, remaining in the hands of the butchers, tanners and horn-workers. It may be therefore that builders and other persons wishing to make use of these bones as building material would have been able to obtain large quantities for very little cost or even free of charge perhaps, providing they were willing to arrange for the transportation of the material.

Unfortunately there is no available information on the costs (if any) incurred by the builders in London who used animal bones. However, there is at least one documented instance from Harrogate, Yorkshire, where a builder carrying out work on a local church roof was charged 3s for 1,000 sheep "shanks" (tibiae) which he was using in lieu of wooden or iron pins to fasten the stone roofing slates to rafters (see Jackson, 1966 and Gazetteer, section 3.3).

Irrespective of whether the waste bone material was obtained free or had to be purchased, the eventual use of this material for building purposes falls into two distinct categories: 1) casual opportunistic use and 2) planned use.

1) **Casual/opportunistic use:**

This category is best exemplified by the actions of the builders who were responsible for repairing the damaged cottage wall in Ware, Hertfordshire (figures 1 and 2) (Gazetteer, section 3.2). Here the builders were clearly merely exploiting in an opportunistic way a convenient source of cheap - perhaps even free - substitute building material provided by the nearby slaughterhouse. Likewise, the builders of the horn core pits in the east end of London (see Gazetteer, section 2.1) were taking advantage of the ready-to-hand cattle horn cores accumulated by the horn-workers in Petticoat Lane and by the builders along Aldgate High Street.

2) **Planned use:**

i) Constructional:

The situations described above may be compared with the planned (premeditated) exploitation by the farmers who constructed the land drains on the Forest House Estate in Essex and at Bows Farm in Middlesex (see Gazetteer, section 5.1). In both of these cases, the farmers must have anticipated using cattle horn cores as a substitute for the more traditional

Number of cattle slaughtered in the City

in one year (note (a))	approx.	88,000
minus no. very young calves (note (b))	approx.	44,000
		44,000 animals
No. metapodial bones produced by these adult cattle		176,000
No. horn cores produced by the cattle killed in London (after allowance has been made for calves and polled cattle) (note (c))	- about	78,000

(a) The number of cattle slaughtered is based on the following sources:
  i) The numbers of cattle sold in Smithfield market each year from 1732 to 1740 given in M'Culloch's Dictionary of Commerce (quoted in Youatt, 1834: 256).
  ii) The Guildhall Record Office Journal of the Markets Committee 1696-1698 fo. 17 which shows that an estimated 2,400 cattle were brought to Smithfield each week in 1696 for sale to butchers (not all of these were destined for immediate slaughter, however, and many were held back for resale at a later date when prices had risen).
  iii) John Houghton's estimate of 88,400 cattle killed in London in 1692 based on information supplied by butchers in the City (quoted in Fussell & Goodman, 1935: 214).

(b) Most sources indicate that of the total number of cattle killed in any one year in London, about half of these were calves (mostly veal calves under one year of age). The bones of these calves would have been too small and fragile for use as building material.

(c) The horn cores of calves would have been very little developed and these animals have therefore been excluded from the calculation. Polled (hornless) cattle are known to have been reared in Suffolk in the late 17th and early 18th century (Kerridge, 1968: 317) but the exact numbers of these animals sent to London is not recorded. It is known, however, that the stock of polled cattle was small compared with the numbers of medium and long horned cattle and that as they were mainly kept as dairy cattle very few were sent to the London meat markets. An arbitrary figure of 5,000 cattle has therefore been adopted to allow for this group of animals.

Table 1. Calculation of numbers of metapodials and horn cores produced by cattle killed in the City of London in one year in the late seventeenth/early eighteenth century.

brushwood to line their drains as their nearest supply of sufficient quantities of suitable material lay some miles distant from their farms, in London - and special provision must have been made for the cattle horn cores to be transported out of the City (see Armitage, Coxshall & Ivens, 1980).

ii) Decorative:

During the late 17th and early 18th century the building of so-called "knuckle-bone" floors became fashionable, especially in Oxford (see Gazetteer, section 1.2). Owners of buildings with simple earth floors would consolidate and protect them from wear as well as producing a pleasing decorative effect - by driving animal bones into the surface. Such floors became known as "knuckle-bone floors" after the bone elements most frequently used in their construction: metapodial bones of sheep and cattle. The term "knuckle-bone" is today more generally applied to cattle and sheep astragali (vide: the game of "knuckle-bones"), but at that time, however, the term referred to the metapodials of animals, as evidenced by Kalm's identification of "knuckle-bones" as the bones "...of hourse or ox-legs, such as the boys...[use] in Sweden and Finland.. to make their so-called 'ice-legs' (bone skates) with which they run on the ice" (Kalm, 1748, reptd. 1892: 67).

Further examples of the planned use of animal bones in the contemporary architecture of this period are still extant today: as can be seen in the incorporation of animal bones as part of the decoration to "follies" built in the grounds of large country estates - here the bones presumably were intended to enhance the bizarre appearance of these structures.

Although the use of animal bones was clearly fashionable in certain circles, the practice did not, however, always meet with approval - as evidenced by Parkinson's observations regarding the construction of fences around flower beds using sheep metapodial bones "...although these will last long in forme and order, yet because they are but bones many dislike them, and indeed I know but few that use them" (Parkinson, 1629). Parkinson also decried the continental practice of using "the jaw bones of asses" for the same purpose, on the grounds that this was being "too gross and base" (see Hurst, 1982: 38).

CONCLUSIONS

From the foregoing survey it is evident that the practice of building in animal bone in the post-medieval period was more widespread than has hitherto been believed. Indeed, there is accumulating evidence that the practice was not confined to Britain but was world-wide. It is beyond the scope of this paper to attempt a review of this foreign evidence, but three examples will suffice to illustrate this point:

i) An article in L'Edis Republicain dated Mardi 12 décembre 1978 describes the discovery by workmen in Chartres of a wall to a bakery packed with cattle horn cores. According to the writer, this wall was thought to date from the 16th century. This find was being investigated by Dominique Joly of the archaeological section, Museum of Chartres (Joly, 1983, pers. comm.)

ii) Excavations at the Convento de Santo Domingo in Panama City uncovered part of a 17th century cobble stone floor incorporating cattle metapodial bones as decoration (Cooke & Rovira, 1983: 54; Rovira, 1984, pers. comm.).

iii) W.H. Hudson in his book Far Away and Long Ago (quoted in Dobie, 1943: 190) described fences of longhorn cattle skulls enclosing "fine homes at Buenos Aires".

Returning to Britain, it is important to note that building in animal bones ceases abruptly after c. 1750. It could easily be argued that the practice simply went out of fashion; but this would certainly be far too simplistic an explanation. Part of the answer may lie in the increasing use of animal bone as fertiliser from the late 18th century onwards. As discussed by Kerridge (1968: 243) farmers in Hertfordshire and other counties around London at this time bought in refuse (including "sheeps trotters" and other bones) for use as fertiliser; after being calcined the bones were ploughed into the soil. Evidence of this practice comes from the survey of Hertfordshire by Arthur Young (1804 reptd. 1971: 167-8) who records that bones were "much esteemed" by the local farmers, and that "they are considered as best for pastures when burnt; but for arable clay better when only boiled"; other farmers apparently thought that raw bones were "...the most lasting of any manures...they enrich the ground for many years; and then the mechanical effect, in lightening it, will last almost for ever". One farmer (Mr. Parker) interviewed by Young (ibid: 168) revealed the source of these bones: "...They may be got at many great towns where no manufacture of bones is carried on". By the 1840s - 1850s, however, the demand for animal bones for grinding down to make bone-meal fertiliser was so great that the supplies of raw bones from urban slaughteryards were being rapidly exhausted and alternative sources had to be sought - leading to the importation of "enormous quantities of bone .. from continental Europe". (Prummel, 1983: 24). England's demand for this raw material was insatiable and led to the digging up of sub-fossil bones from the early medieval buried refuse dumps at Dorestad in the Netherlands (ibid: 24). Similarly, cattle skulls and other bones from the Roman levels in London were collected and carted away to nearby processing factories for conversion to fertiliser (see Lane Fox, 1867).

On this evidence it seems very likely that the use of animal bones as building material was no longer economic after the mid 18th century as virtually all of the bone waste formerly so readily available in such abundance from cities and towns was being diverted for use by farmers as fertiliser.

Today, few examples have survived to remind us how extensive was the practice of using animal bones as building material; and it is important that these are properly surveyed, documented and protected for the benefit of future generations.

ACKNOWLEDGEMENTS

Much of the research for this paper was carried out while the author was Environmental Research Officer at the Museum of London's Department of Urban Archaeology, and was in part financed by grants issued through the then D.O.E. Ancient Monuments Laboratory (now part of HBMC).

I would like to thank all those friends and colleagues, particularly those from the Museum of London, and British Museum (Natural History) who provided information and helpful advice, and without whose support this paper would not have been possible.

REFERENCES

ARMITAGE, P.L. (1982) Studies on the remains of domestic livestock from Roman, Medieval and Early Modern London: objectives and methods. In A.R. Hall & H.K. Kenward (eds) Environmental Archaeology in the Urban Context. Council for British Archaeology Research Report No. 43: 94-106.

ARMITAGE, P.L., COXSHALL, R. & IVENS, J. (1980) Early agricultural land drains in the former parishes of Edmonton and Enfield. The London Archaeologist, Vol. 3 (No. 15): 408-415.

BUCHANON, R.H. (1956) A buried horse skull. Ulster Folklife, vol. 2: 60-61.

BURDICK, L.D. (1901) Foundation Rites. London: Abbey Press.

COOKE, R.G. & ROVIRA, B.E. (1983) Historical archaeology in Panama City. Archaeology, vol. 36 (No. 2): 51-57.

CROSSBY, F. (1974) Ware. Hertfordshire Archaeological Review, No. 9: 175.

DOBIE, J.F. (1943) The Longhorns. London: Nicholson & Watson.

EVANS, G.E. (1966) The Pattern Under the Plough: Aspects of the Folk-Life of East Anglia. London: Faber & Faber.

FUSSELL, G.E. & GOODMAN, C. (1936) Eighteenth century traffic in livestock. Economic History, vol. 3 (Part 11): 214-236.

GAILEY, A. (1971) Horse skulls under a County Down farmhouse floor. Ulster Folk Museum and Transport Museum Year Book 1969-70: 13-14.

HURST, R. (1982) Seventeenth century garden recreated at Capal Manor. Hertfordshire Countryside, vol. 37 (No. 279): 38-39.

JACKSON, S. (1966) Sheep shanks. Bradford City Art Gallery and Museums Archaeology Group Bulletin, vol. 11 (No. 4): 1.

KALM, P. (1892) Visit to England (translated from the Swedish by J. Lucas from 1748 edn.).

KERRIDGE, E. (1968) The Agricultural Revolution. New York: Augustus M. Kelley.

LANE FOX, A. (1867) A description of certain piles found near London Wall and Southwark, possibly the remains of pile dwellings. Journal of the Anthropological Society of London, 5: 71-80.

MERRIFIELD, R. (1969) Folk-lore in London Archaeology Part 2: the post-Roman period. The London Archaeologist, vol. 1 (No. 5): 99-104.

PARKINSON, J. (1629) Paradisi in Sole Paradisus Terrestris. London.

PORTER, E. (1969) Cambridgeshire Customs and Folklore. London: Routledge & Kegan Paul.

PRUMMEL, W. (1983) Excavations at Dorestad 2 Early Medieval Dorestad an Archaeozoological Study. Amersfoort ROB.

SANDKLEF, A. (1949) Singing Flails. A Study in Threshing-floor Constructions, Flail-threshing Traditions and the Magic Guarding of the House. Folklore Fellows Communications No. 136: Helsinki.

SCHOFIELD, J. & DYSON, A. (1980) Archaeology of the City of London. London: Museum of London & Mobil.

SPETH, G.W. (1893) Builder Rites and Ceremonies. Reprint of two lectures given at the Church Institute Margate. MS held by London University Library.

SUILLEABHAIN, S.O. (1945) Foundation sacrifices. Journal of the Royal Society of Antiquaries of Ireland, vol. 75: 45-52.

YOUATT, W. (1834) Cattle: their Breeds, Management and Diseases. London: Robert Baldwin.

YOUNG, A. (1804 reptd. 1971) General View of the Agriculture of Hertfordshire. Newton Abbot: David & Charles.

BONE ANALYSIS AND URBAN ECONOMY: EXAMPLES OF SELECTIVITY AND A CASE FOR COMPARISON

Bruce Levitan

INTRODUCTION

Bone analysis from urban sites has, in recent years, become increasingly common, mainly as a result of the "Rescue boom" of the early 1970s. Some idea of the scale of this can be deduced from the various contributions to the recent volumes produced by DoE/HBMC on the subject of environmental archaeology, e.g. Armitage et al (1987), Bell (1984), Coy and Maltby (1987) and Kenward et al (1984).

One result·of this has been the publication of major studies of bone assemblages from urban sites, e.g. Maltby (1979) for Exeter, Noddle (1985) for Hereford, and O'Connor (1982) for Lincoln. Such studies have taken bone analysis beyond the stage of analysis by rote, and, as well as including much research·and original thought, have become advocates of increasing selectivity. This is partly because the assemblages dealt with have been very large, e.g. 75,000, 14,000 and 60,000 bones respectively for the sites quoted above. Dealing with such large quantities of bone has led to the recognition that certain patternings within the evidence are repeated on different sites, and mere replication of information is of little use or sense. Furthermore, when the total number of bones from many sites from one town is considered, the size of the problem of analysis can be staggering: O'Connor (1984, 3) estimates more than three million bones for York.

It is with these points in mind that this paper is written. The aim is to show that a selective approach to bone analysis may not only be more cost effective, but can focus on specific problems in an analysis that is uncluttered by other, more routine aspects. In each case a different aspect of the bone assemblage is considered, the intention being to show that the essential characteristics of the sites can be thus highlighted. In addition, the aim is to show that the analysis should be integrated with the archaeology of the site, and that the bone report should be more than a mere appendix of the site report.

The comparative approach is dealt with only briefly because three sites are not a good enough basis for a case study of inter-site comparison. It is a corollary, however, of the selective approach, since it represents the next stage of analysis beyond the site report: that of inter-site comparisons from one town. It seems superfluous to have to suggest that this is a necessary step, yet the approach to excavation and analysis from urban sites is often too limited in scope, almost as if there is a "dig it because it's there" reasoning, rather than part of a problem-orientated strategy for the town as a whole, and there is an obvious knock-on effect for bone anlaysis. That bone analysis has suffered from this situation is evidenced by the recent review of evidence from medieval sites in South West England by the author. Despite the existence of over 40 bone reports from towns in the region, it was impossible, except in the case of Exeter, to make any useful generalisations about the information produced (Levitan, 1987).

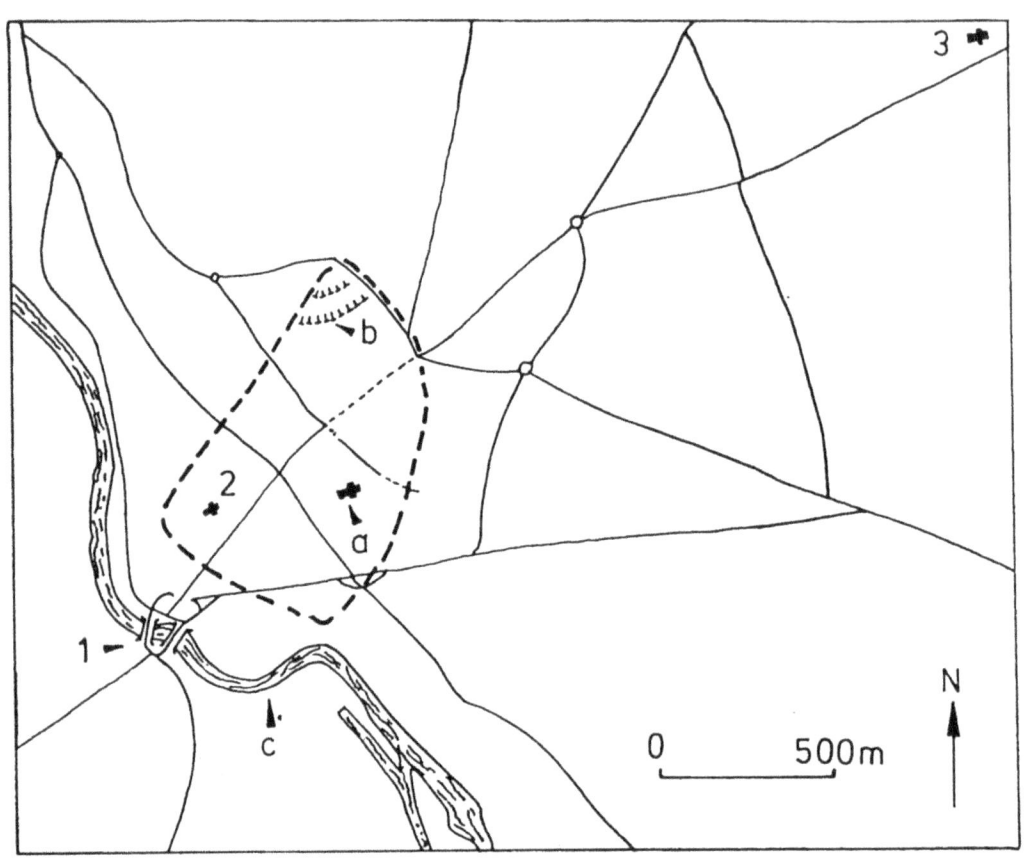

Figure 1. Exeter: location of sites discussed.

EXE BRIDGE: BODY PART REPRESENTATION AND SITE FUNCTION

The excavations at Exe Bridge, Exeter, (1975-1979) were directed by Stuart Brown for the Exeter Museums Archaeological Field Unit. The site was on the east bank of the River Exe, on the north side of the bridge and St. Edmund's Church (Figure 1). The Exe Bridge was constructed c 1200 AD, and replaced an earlier timber bridge (Allan, 1984, 59-67; Henderson, 1985, 6). The excavations included parts of St. Edmund's Church which was at the northern end of the bridge, but the bones were mainly recovered from three tenement plots adjacent to the church and bridge.

A total of 12163 bones were examined, and most of them (6439) come from one phase dated to the thirteenth century. Other phases total between 695 and 2080 bones. The major group of animals represented is the mammals, with 9510 bones, and over 90% of the bones identified to species level comprise cattle and sheep/goat. This is an unusually high percentage, though cattle and sheep/goat are generally the most important animals represented at other sites from Exeter (Maltby, 1979).

The best evidence for site function here, however, comes from an analysis of body-part representation, and not from species representation. Before presenting the results of body-part representation, a brief discussion of the methods of analysis is necessary. The use of simple fragment counts is not the most appropriate method for this analysis for three reasons. Firstly, the skeletons of different animals have different numbers of certain elements. Secondly, and related to the first point, different elements within a single skeleton occur in different numbers. Thus, in a fragment count, the relative frequency of the elements may be biased. The third factor is fragmentation. The bones will be fragmented to varying degrees, and this may depend upon a number of factors which can act separately or in combination, for example butchery, cooking, scavenging, trampling. It follows that more fragile bones (e.g. skull) may be more greatly fragmented than stronger bones (e.g. phalanges), and some bones may be more heavily fragmented due to butchery (e.g. upper leg bones which might have been smashed open for marrow extraction) than others (e.g. vertebrae which may only have been chopped in half when dividing the carcase). It should be clear from this that these three biases make it impossible to assess the evidence of anatomical representation from fragment counts alone.

There are several ways of combating these biases (e.g. Grant, 1975; Watson, 1979; Levitan, forthcoming a), and the method employed here is to standardise for expected anatomical frequency, and to discount very small fragments (i.e. those less than a quarter complete).

The results for cattle and sheep/goat are given in Tables 1 and 2. The column headed N gives the results of the bone counts with fragments less than a quarter complete excluded. The N/EF column gives the results when N is standardised for skeletal frequency. What is immediately evident from the tables is that horncores dominate the deposits. For the thirteenth century, in the case of cattle, 450 out of a total of 759 bones are horncores, and for sheep/goat there are 467 horncores out of a total of 1332 bones. This pattern is repeated in the later periods, though to a lesser extreme, and it should be noted that the sample size also decreased in later periods.

In the thirteenth century, goats account for 422 (94%) of the sheep/goat horncores, yet on the basis of other parts of anatomy, sheep outnumber goats by 2:1 (184 bones identified as sheep and 95 as goat).

Figure 2 illustrates the anatomical representation of cattle and sheep/goat (labelled sheep in the figure) for the thirteenth century. Besides the emphasis on horncores, a secondary emphasis on metapodials (lower limb longbones) is evident for sheep/goat. Note that here sheep are dominant with 106 metapodials compared with 40 of goat (a further 220 could not be assigned to species).

Clearly, this deposit, comprising such high percentages of horncores, represents some specialised process. In fact, two activities appear to be represented. Interpretation of one is partly based upon evidence for temporal differences in quantity of bones. The clear concentration of bones in the thirteenth century represents a peak of bone deposition at the time, and this fits in well with the rest of the archaeological evidence which indicates reclamation of the river bank by dumping of large quantities of bones and other material. This phase of reclamation ceased in later periods, and two tenement properties were extended onto the new ground. Such deposits, where river bank reclamation comprises the dumping of bones and other rubbish, have many parallels in other cities, e.g. London, Bristol and Ipswich.

The fact that the deposits comprise mainly horncores is not relevant in terms of bones chosen for reclamation, but it is important in that the bones were presumably gathered from close by. The postulated proximity of the activities which generated the deposits is further attested by the fact that horncores remain an important component in later periods. On a note of caution, however, much of the late thirteenth-fourteenth century pottery in Tenement A is residual. This may be true to a lesser extent in Tenement B (Allan, 1984, 59). These deposits probably relate to hornworking, bone working and/or tanning (see MacGregor, this volume; Serjeantson, this volume).

The last of these would be characterised by high proportions of horncores and limb extremity bones since these were generally detached with the hides. Tables 1 and 2 and Figure 2 show relatively high proportions of limb extremities (metapodials and phalanges). The presence of metapodials might also argue for bone working since this element was often used. The lack of butchery evidence on the metapodials, however, argues against this, since the residue from metapodials used for bone working commonly consists of proximal and distal ends only, the shafts having been used for bone working (Armitage, 1982a, 104; MacGregor, 1985, 47).

Hornworking is the activity that best fits the evidence, and similar patterns have been interpreted as hornworking waste at other sites, e.g. Augst, Switzerland (Schmid, 1968), Angel Court, Aldate and Cutler Street, London (Armitage, 1982a). Other sites are quoted by MacGregor (1985, 51-53). It is ironic that Maltby (1979, 86) noted an <u>absence</u> of horncores in second and fourth century deposits from Exeter. He suggested that the under-representation of horncores implied reservation of horncores elsewhere in the city. MacGregor (1985, 51-53) notes the characteristic butchery whereby cores are hacked and broken from the skull for hornworking or tanning, and similar patterns of butchery were recorded here. A recent find of horncore deposits from Exeter that possibly relates to tanning rather than hornworking was reported by Levitan (1985).

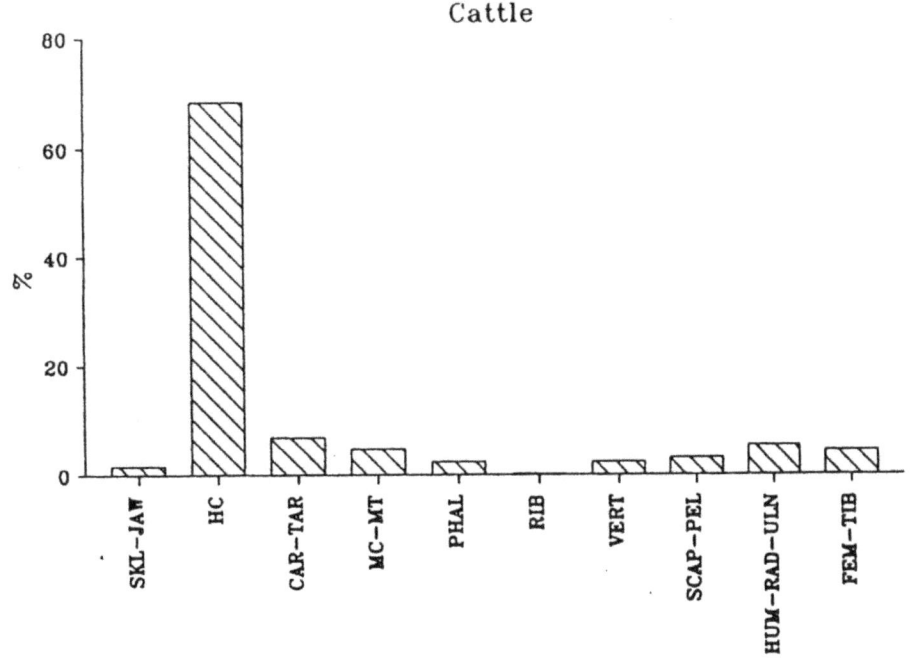

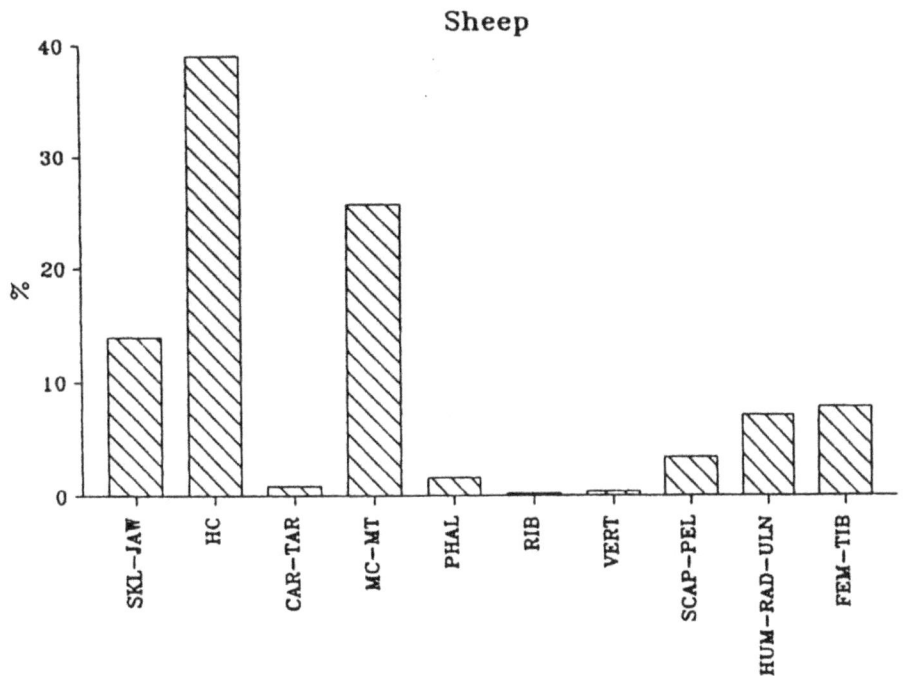

For key to anatomical groups see Table 8

Figure 2. Exe Bridge: anatomical groups.

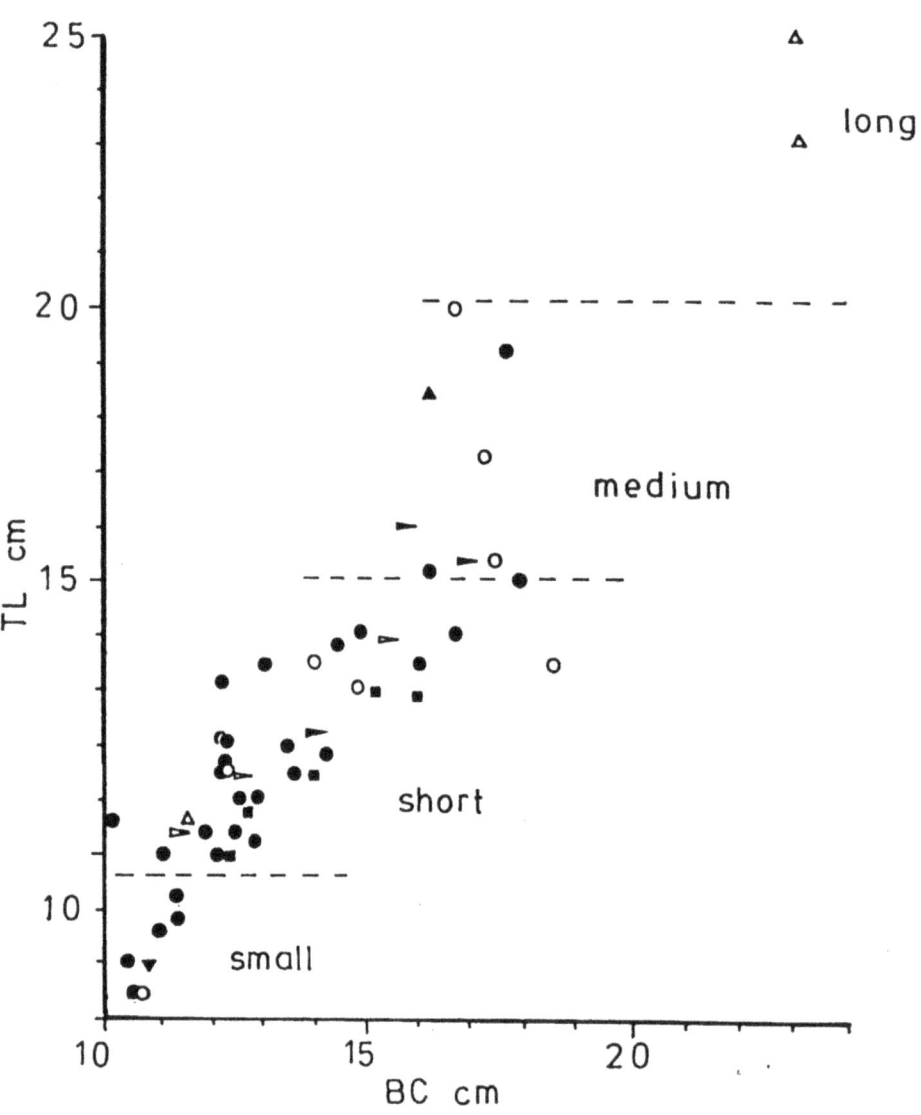

Figure 3. Exe Bridge: scattergram of cattle horn core measurements (cm). TL = total length; BC = basal circumference.
The four size groups are those defined by Armitage (1982b).

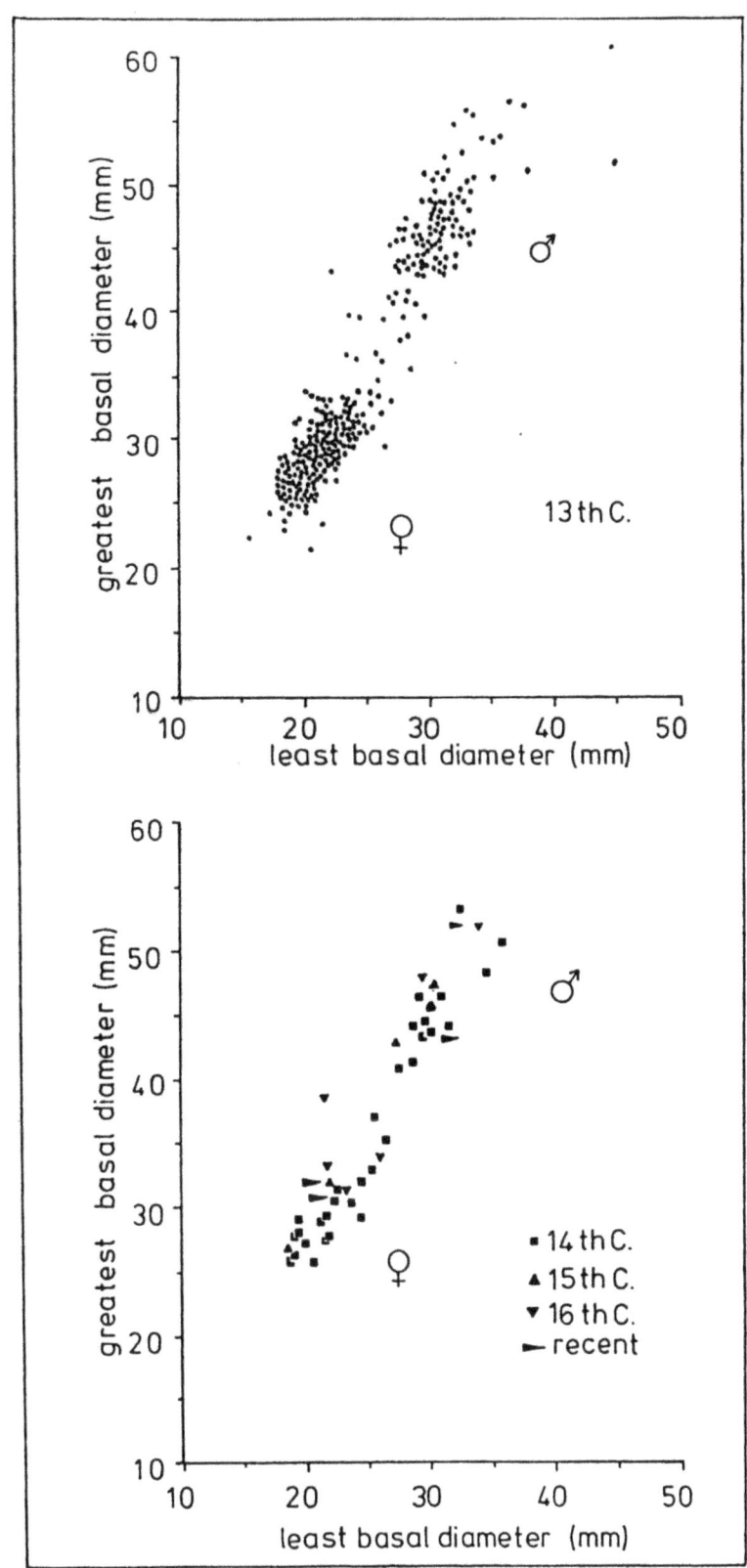

Figure 4. Exe Bridge: scatter plot of goat horn core measurements (mm). Greatest basal diameter is plotted against least basal diameter. Top: thirteenth century. Bottom: fourteenth century to recent.

The horncores can be used to gain insights into age profiles of cattle (Armitage, 1982b) which can be compared with those from other bones. The 330 aged horncores are summarised in Table 3 which shows a range of ages from less than a year to over ten years old. The majority are from animals aged three to seven years old, so presumably older animals were preferred (with better developed horns). The evidence from fusion of the bones (Table 4) implies that the majority of cattle were killed at around three years old. Market forces would undoubtedly have favoured younger cattle since meat would have been the most important product from cattle sold in Exeter, so the hornworkers' raw material of young adult horncores reflects this effect.

Sex determination can sometimes be carried out using metrical evidence. In the case of horncores (where there are large samples of measured bones) this is less useful in the case of the cattle, where there is a lot of non-sex based variation, (Figure 3) than for the goats (Figure 4). The cattle evidence is complicated by the presence of different size classes, and these are defined by Armitage and Clutton-Brock (1976) and Armitage (1982b). Although these results may not be informative in terms of sex separation, they do show that the short horned variety of cattle was the commonest, providing useful information about the main type of cattle was exploited. In the case of goat horncores there is a good correlation between basal dimensions and greatest length (Levitan, forthcoming b), and whilst the latter give better sex separation results, the former have been employed here due to larger sample size (Figure 4). The larger horncores, in the top right part of the scatter plots, are probably males. Males are slightly less common than females, but the difference is not great. The implication, for goats at least, is that there was no great preference in terms of sex, for horn raw material.

ST. KATHERINE'S PRIORY: A STUDY OF LATERAL VARIATION

St. Katherine's Priory (Figure 1), a Benedictine nunnery dating from 1160 AD to 1538 AD, was excavated between 1976 and 1978 under the direction of John Allan. The excavations were mainly concerned with the cloisters and associated buildings to the south of the church, though nearly all the church was also uncovered in the north part of the site (Figure 5). The main deposits, 7065 out of 10197 bones, relate to the period c 1500-1530 AD, with bones from other periods totalling 314 to 957. The majority of bones are from the southern part of the site, and these could be divided up according to location, thus allowing a study and comparison of the lateral variation within these deposits. The analysis of lateral variation here is a useful tool in helping to provide information about the activities in the different locations, and similar studies have been used to good effect on different types and periods of site by Wilson (1978; 1985).

One way of looking at the lateral variation is in terms of species representation plotted as time series graphs (Figure 6). It is very clear from this figure that each location gives a different view of the relative abundance of the species through time. The kitchen areas (interior and exterior) indicate an increase in the importance of cattle, set against a decrease in sheep/goat. In the dorter area, sheep/goat are much more numerous in all periods, and there are decreases in sheep/goat and cattle through time. In room A, sheep are dominant, but an overall fall in percentages is set against a rise in cattle percentages. The garden deposit shows a very variable picture in the relative proportions of sheep and

1-Church; 2-West Range; 3-Cloisters (room A); 4-Chapter House; 5-Kitchens; 6-Refectory; 7-Dorter; 8-Garden; 9-Kitch. exterior.

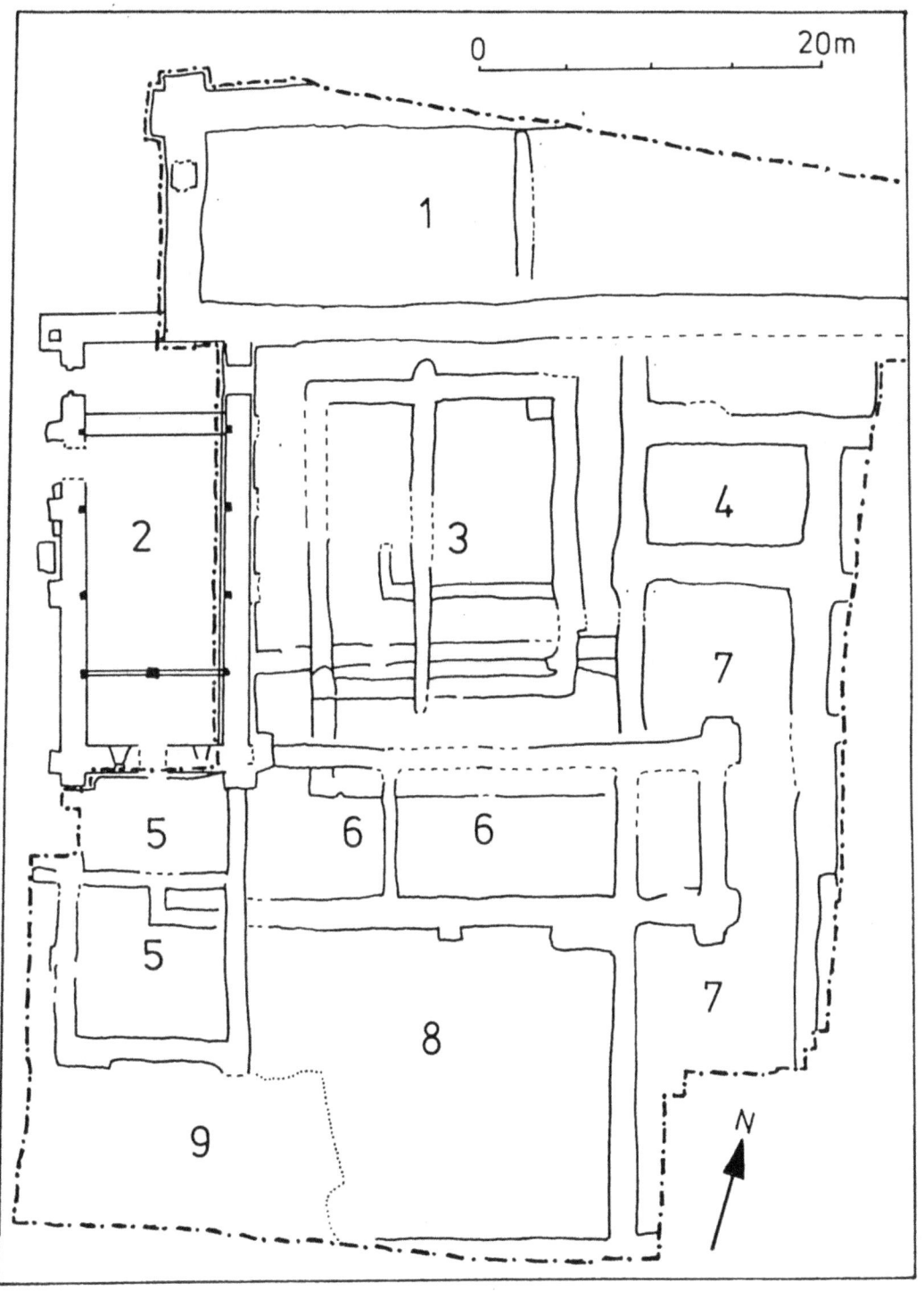

Figure 5. St. Katherine's Priory: plan.

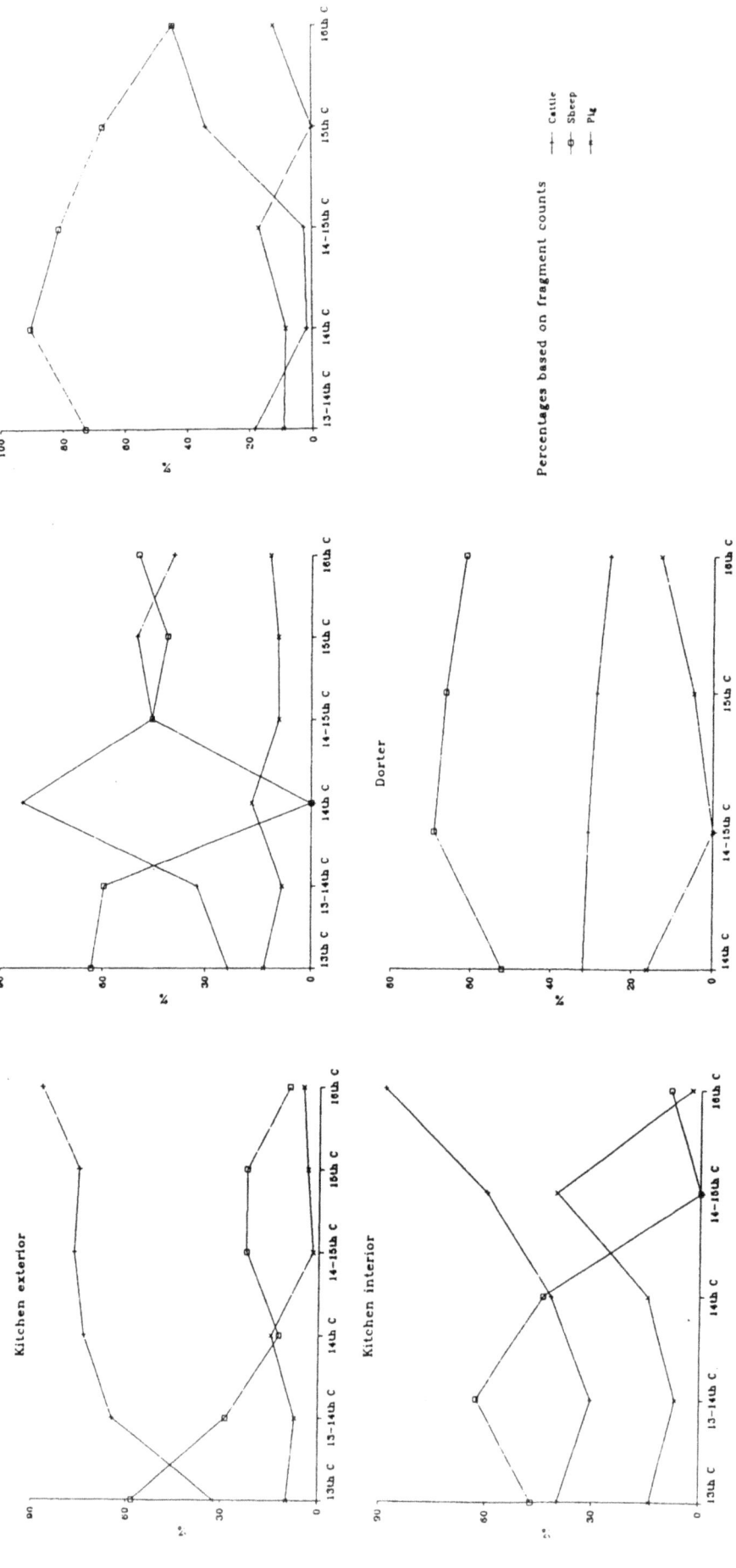

Figure 6. St. Katherine's Priory: graph showing percentages of the major mammals in five locations in the priory.

cattle through time. These results are important in two senses. Firstly, they indicate that the bone deposits are not uniform, so that a simple combination of all the deposits might not give a representative view of the exploitation patterns. Secondly, they show that the different deposits change through time, so they cannot necessarily be seen as temporally consistent.

Another way in which the lateral variation can be investigated is in terms of distribution plots. Wilson (1985; Wilson and Levitan, forthcoming) has recently shown that Iron Age sites may display variation in deposits in terms of size of bones. He did this by plots of bones from smaller animals (sheep, goat and pig) expressed as a proportion of all bones (i.e. small and large bones; the latter being represented by cattle and horse). The variation here has been investigated by similar means (Figure 7). It must be admitted that this is a crude estimate since the cattle and horse bones will comprise some small fragments, and the sheep, goat and pig may have some complete large bones. Without the help of a computer, however (this site was analysed manually), a plot which also takes fragmentation into account would be very difficult to do with a large sample. The biases of including some bones in the wrong categories are not thought to be too great, so the results should still be reasonably informative.

The main pattern that emerges is that the smallest proportions of small bones occur in the kitchens exterior (in all periods except the fourteenth century), where the proportions are the second smallest. This indicates that the largest bones were consistently deposited in the kitchens exterior area. The greatest accumulation of small bones occur within the dorter and room A, and also occasionally in the garden (thirteenth-fourteenth centuries and sixteenth century). The kitchens are the most variable, with mainly small bones in some periods (especially thirteenth-fourteenth centuries) and mainly large bones in other periods (e.g. sixteenth century).

These results show that a reasonably clear pattern occurs in all periods, though it is best exemplified in the fifteenth and sixteenth centuries. The large bones represent waste from secondary butchery (and possibly even primary butchery for pigs) in kitchen preparation. This would involve complete or half carcases which are cut into joints and boned out in the case of cattle in particular, with the bones being waste at this stage. These bones, then, were dumped in deposits near the kitchen, in ditches which originally served for water supply, and were presumably used as a convenient area for rubbish dumping when they went out of use. The small bones, which represent table waste, are in deposits closer to the eating and living areas, and were perhaps disposed of more haphazardly (which may account for the variation in deposits in the kitchen and garden, for example).

It would appear, therefore, that the priory was buying in carcases whole and/or halved, and a lot of secondary butchery occurred on site. This is supported by the anatomical representation (Tables 5 and 6) which includes typical butchery waste such as limb extremities and skulls, and, in the case of cattle, some long bones as well.

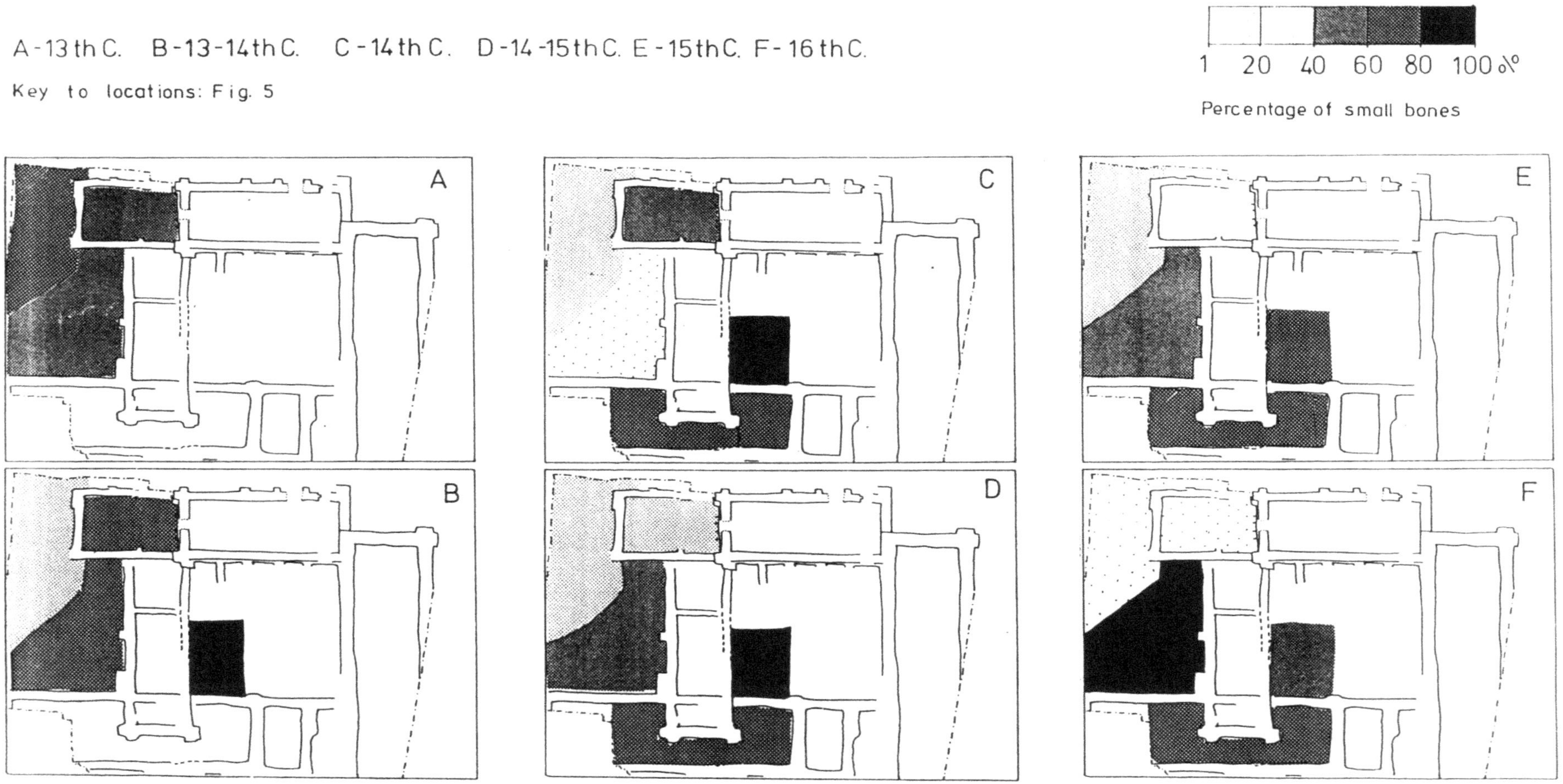
A-13thC. B-13-14thC. C-14thC. D-14-15thC. E-15thC. F-16thC.
Key to locations: Fig. 5
1 20 40 60 80 100%
Percentage of small bones
A
B
C
D
E
F

Figure 7. St. Katherine's Priory: percentage of small bones.

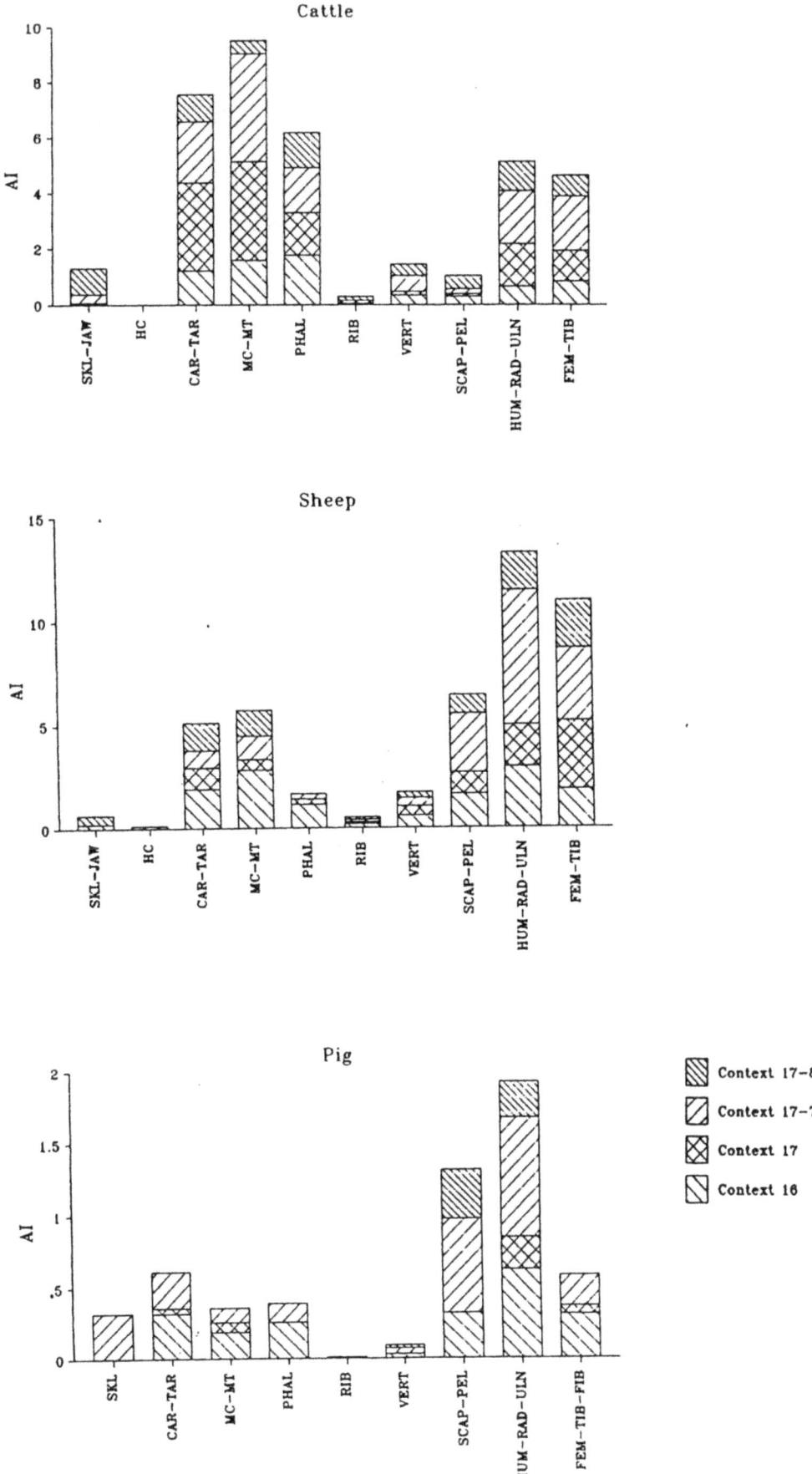

Figure 8. St. Katherine's Priory: histogram of anatomical groups from four contexts.

ST. NICHOLAS PRIORY: TEMPORAL PRECISION

This Benedictine priory (Figure 1) was founded during the reign of William I and dissolved in 1536. It was the richest of Exeter's monasteries, and after the Dissolution became a grand town house, later succeeded by poorer housing (Allan and Henderson, 1984).

Occasionally bones may be recovered from clearly dated contexts of a short time span, and, if luck prevails, the bone groups may be large enough for detailed analysis. The 1983-84 excavation at St. Nicholas Priory, directed by John Allan, provided such an assemblage, with three groups of bone from back-filled robbing trenches of the church and tower which date to 1536-1550 AD, and a fourth group dating perhaps twenty or so years later. These yielded a sample of 4939 bones of which 3870 were identified (Table 7). This provides an opportunity to look at a series of rubbish dumps in detail; from which information might be elucidated about the make-up of the dumps. Are they specific types of rubbish, e.g. primary butchery, domestic? The only major problem in dealing with this group is the lack of sieved bones, so small bones, especially those of fish, will be under-represented (c.f. Coy, this volume). Bulk samples were taken for sieving, but this had not been completed at the time of the analysis.

Table 7 summarises the animals represented, and the first thing to note is that three species account for about 90% of the identified bones: cattle and sheep/goat. Only five of 282 sheep/goat bones identified to species level are goat (the rest being sheep), so it is safe to assume that cattle and sheep were the main species. The second point is that there is a range of mammal, bird and fish species represented: twelve species of mammal, at least nine of bird and a minimum of six species of fish. It is worth reiterating the under-representation of small bones (birds and fish), and the number of fish species could probably be more than doubled if the bones from bulk sieving had been included.

One question that arises from this assemblage is: Is it possible to calculate how many animals were consumed? If so, this will give some indication of the importance of the various species in the diet. Finding the answer is extremely difficult. Even supposing that accurate numbers of animals can be calculated (one possible way being the minimum number of individuals (MNI) method), it is much too simplistic to extrapolate from such a result. This is because it is unlikely that whole animals were bought and consumed. It is much more likely that they were consumed as prepared joints, so the various parts of single individuals might become widely dispersed. For example, some parts of the body may be detached and disposed of as waste at the primary butchery stage, such as parts of the skull. Some other bones extracted during primary butchery may be sold for bone working or left on hides sold for tanning and horn working. A second group of waste bones may result from secondary butchery, i.e. preparing the meat for retail or cooking. Finally, joints of meat will be sold, some with bones in the joint, and these will become part of the domestic rubbish dumps. The larger the animal, the more complicated the stages of butchery and bone removal, so cattle will be subjected to more butchery than pigs and sheep, and small mammals, birds and fish may be sold as complete carcases.

What has been attempted is the identification of specific types of rubbish deposit in terms of different cuts of meat consumed for the major meat producing animals: cattle, sheep and pig. The counts for each skeletal part have been transformed into an anatomical index (AI) which standardises for different parts of the skeleton and for different degrees of fragmentation. This makes each index directly comparable with any other from any part of the body and from any species. The method is fully described in Levitan (forthcoming a). The index provides only a crude estimate of frequency since the AI (see Table 8) ignores fragments less than 25% complete, and such fragments frequently make up the bulk of the identified assemblage. The 25% column in Table 8 shows the proportion of these fragments out of the totals for each anatomical element and there is clearly much variation, with the more fragmented bones being highlighted by high percentages in this column.

Table 8 also lists the ranks of the 29 anatomical groups. Interestingly, the prime meat producing bones from sheep/goat occupy three of the top five ranks. This could imply that the rubbish consists of domestic waste since the relative paucity of the same anatomical parts of cattle may result from boning out at an earlier stage in the butchery process, and there would be little domestic bone waste from cattle. It is surprising, therefore, to find that the best represented cattle bones are carpals, tarsals and metapodials (from the lower part of the limb). This, however, may be another reflection of greater fragmentation of meat bearing bones of cattle compared with those of sheep due to butchery. The table shows that between 62% and 89% of the cattle bones are in the 25% category, compared with values of 17%-45% for sheep and 20%-43% for pig. This still does not fully account for the fact that cattle metapodials are so well reperesented (second in rank of all the anatomical groups, Table 8). A secondary component of the rubbish, therefore, might be waste from secondary butchery of cattle.

The table, which shows the relative frequency of the different anatomical groups, indicates that there is a mixture of deposits by highlighting the high incidence of meat-joint bones of sheep (which are the most frequent) and the butchery waste component of cattle. One of the advantages, described above, of analysing a deposit such as this, is that one can look at variations within the deposit. Might it be that the two components described above (i.e. domestic waste and secondary butchery waste) can be recognised in different parts of the deposit?

Four main sub-groups within the deposit are summarised in the table, and these are represented graphically in Figures 8 and 9. These sub-groups have been selected on the basis of sample size, the other sub-groups being too small for anatomical analysis.

Figure 8 (top) shows the results for cattle bones, in terms of AI counts, and clearly there are differences for some of the anatomical groups. The main differences are for the carpals/tarsals, metacarpal/metatarsal, humerus/radius/ulna and femur/tibia (i.e. the limb bones). Context 17-7 is characterised by particularly high proportions of upper limb bones, and also by high proportions of the lower limb bones. Context 17 is similar in the latter respect, but has much lower proportions of limb bones than the other contexts. Context 17-1, therefore, appears to represent the greatest concentration of butchery waste, with upper limb bones being those which were removed from joints before retail, and lower limb bones representing secondary butchery waste. The relatively small proportions (in all the contexts) of skull, horncore, mandible and phalanges implies that primary butchery waste is not present. Ribs and vertebrae, which might have been part of domestic waste to a greater extent than the limb bones, are also poorly represented, so this indication is of retail butchery waste. Context 17 also resembles this pattern, though the emphasis on lower limb bones indicates a greater amount of pre-retail butchery waste. The other contexts, with much lower frequencies of bone, are more likely to be domestic waste.

Figure 8 (middle) illustrates the results for sheep. Here there is a very obvious difference from the cattle results, with relatively lower frequencies of lower limb bones, and higher frequencies of upper limb bones. Context 16 bears the closest resemblance to the cattle results, with parallels in the same context. In this respect it is unlike the other contexts for sheep, with higher frequencies of lower limb bones and equal frequencies of upper limb bones. Apart from the almost complete absence of cranial elements in this context, the pattern resembles one where all parts of the butchery process are represented. Table 8 reveals that 87% of the cranial material is less than a quarter complete, so it might well be that all of the butchery processes are represented (note also that the high percentage of ribs less than a quarter complete which may compensate for the low frequencies in the figure). The other three contexts are all alike in terms of lower limbs, but context 17-7 has very high frequencies of upper limb compared with the other contexts (particularly of upper fore-limb). Note, however, that 66% of upper hind limb are less than a quarter complete, the highest for the contexts, so there are also high frequencies of these elements. In this respect, the context 7-17 results may indicate kitchen waste, though a more detailed analysis would have to be undertaken to check this. For example, one would expect to find lower frequencies of distal radius since this part of the bone may have been removed by the butcher. A complicating factor is the fact that the figure shows apparently few cranial remains (a few horncores only), but Table 8 shows that all such remains (excluding horncores) are less than a quarter complete. Thus possibly the deposit is more of a mixture than the figure implies. The two other contexts are essentially similar, and indicate a similar kind of deposit to context 17-7, though less extreme in the representation of upper limbs.

The pig results are illustrated in Figure 8 (bottom) for completeness, though the small samples (Table 8) render those results rather unreliable. In general terms, contexts 17 and 17-8 form a similar pair. Upper limbs are better represented than the other body parts.

The A1 results, illustrated in Figure 8, are shown as cumulative percentages in Figure 9. This method of illustrating the results shows differences in representation by changes in the slope of the curve. For example, for sheep (Figure 9, middle), it is evident that the steep lines for girdles and upper limbs are different to the gently sloped lines for axial bones and lower limbs. This implies that the girdles/upper limbs are better represented, and the results for context 7-17, described above, are exemplified here by the steepest line. All the deposits are poor in phalanges, ribs and vertebrae, and the contrast between context 17-7 and the others is also well illustrated. Similarly, context 17-8 follows a rather different curve for cattle to the other contexts (Figure 9, top).

The attempt to locate different types of rubbish within the deposits has met with limited success. The cattle bones appear to fall into at least two discernable deposits, exemplified by contexts 17 and 17-7 (primary butchery waste?) and contexts 16 and 17-8 (retail waste?). The sheep bones also appear to form two groups, typified by context 16 (mixed material: all levels of waste?) and the other three contexts (retail waste?). It has been possible to show, however, that different types of deposit are present, and

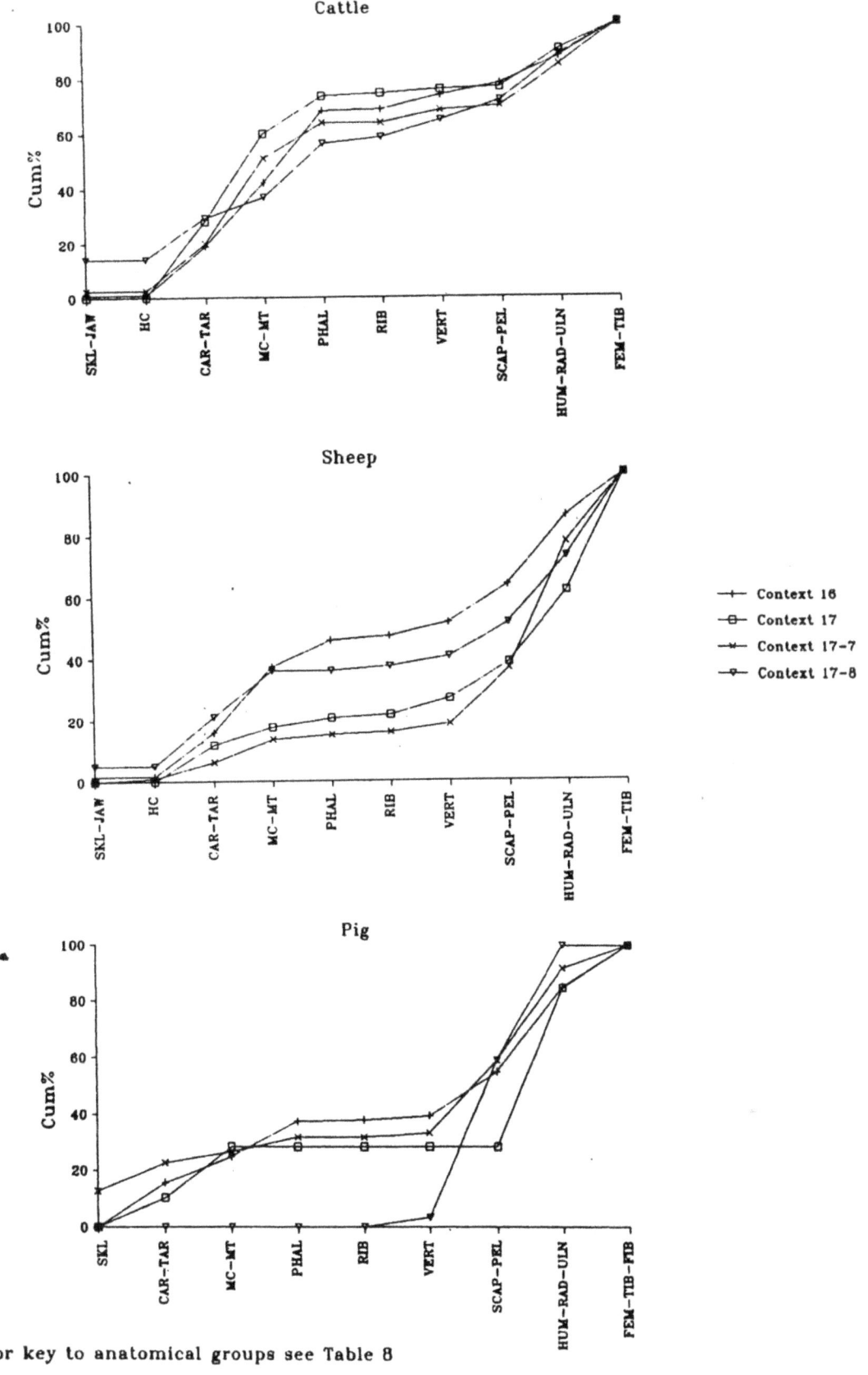

Figure 9. St. Nicholas Priory: cumulative percentages of anatomical groups.

this means that discrete dumps of rubbish are discernable.

CONCLUSIONS: SELECTIVITY AND COMPARISON

The three sites discussed above exemplify the selective approach to bone analysis from urban sites. They have been chosen to illustrate that this approach is a useful tool for both bringing out important aspects of the assemblages and for avoiding the spending of time on repetitious information.

Examples of sites like Exe Bridge are useful in pinpointing information about specialist activities. Such activities, like the one represented at Exe Bridge, are sometimes peripheral to the main meat-based economics of bringing food animals to the city. Comparisons between such sites and sites which are more directly involved with the food-production economy will serve to show how the various animal products activities related in time and space. The Exe Bridge example showed that cattle and sheep/goat horns were utilised at or near the site, and hinted at a relationship between the preferred age at death for horn raw material and the market forces which prevailed: horns from mature individuals were preferred for horn-working, but the market economy for meat limited this supply.

In the second example the results indicated how analysis of lateral variation can give clues about processes and activities within a site. To some extent the comparison of the different deposits within St. Katherine's Priory, both spatially and temporally, could be echoed on the larger, city-wide scale. In another example of this kind of approach, O'Connor illustrated that special deposits within a site can be identified. His analysis of the bones from Caerleon (O'Connor, 1983; 1986) highlighted the "snack-bar" economy of the Roman baths. Further examples of specialisation in Roman marketing activities come from Maltby's report on Exeter (1979, 82-94).

The third example was, perhaps, the least successful, but it helped to highlight the fact that simple questions about site function are often the most difficult to answer. Advances in the methodology of bone analysis are bringing us closer to the means of answering such questions (e.g. O'Connor, this volume), but there is still a long way to go (Bailey and Grigson, 1987). Deposits like the one at St. Nicholas Priory provide us with an opportunity to investigate such questions with problems about archaeological and temporal precision removed. In the relatively superficial analysis employed it was possible to point to discrete rubbish dumps within the main ditch deposit.

It is only at the final, and as yet unexplored, stage of analysis that a good idea of the animal based economy of a town may be gained. This is at the level of inter-site comparison, touched upon above. Not only do we need to compare many such sites as those outlined above from a single town, but sites from the hinterland of the town must also be considered (Levitan, 1987). When one considers the complex inter-relationships that exist between town and country (e.g. O'Connor, 1984) one becomes aware of the futility of pursuing non-question orientated site analysis. Information about what kind of sites should be sought may come not only from bone analysis, but also from sources such as historical documents (Gerrard, 1987). The comparative study by Maltby for Exeter was published a decade ago, yet there are few other reports of its kind. At a time when resources are limited, strategies should be tailored to fit the problems to be tackled as discussed by O'Connor elsewhere in this volume. Excavators must be encouraged to build bone recovery (with bulk sampling and sieving a standard requirement on certain sites) and analysis into their excavation strategies with just as much enthusiasm as they would pottery or other artefacts.

ACKNOWLEDGEMENTS

The bone analyses of three sites discussed above were all funded by the Department of the Environment/HBMC as part of the South West England Faunal Remains Project based at Bristol City Museum and Art Gallery (1981-1986). I am grateful to the excavators, Stuart Brown and John Allan, and to Chris Henderson, Director of the Exeter Museums Archaeological Field Unit, for their help with the selection of the samples, and to John Allan, Chris Henderson and Elspeth Levitan for reading and commenting on this paper.

REFERENCES

ALLAN, J. P. (1984) Medieval and Post-Medieval Finds from Exeter 1970-1980 Exeter: Exeter Museums Archaeological Field Unit (Exeter Archaeological Report 3).

ALLAN, J. P. and HENDERSON, C. G. (1984) St. Nicholas Priory excavations in C.G. Henderson (ed) Archaeology in Exeter 1983/4 Exeter: City Council (Exeter Museums Archaeological Field Unit), 14-17.

ARMITAGE, P. L. (1982a) Studies on the remains of domestic livestock from Roman, medieval and early modern London: objectives and methods in A.R. Hall and H.K. Kenward (eds) Environmental Archaeology in the Urban Context London: Council for British Archaeology (Research Report 43), 94-106.

ARMITAGE, P. L. (1982b) A system for ageing and sexing the horncores of cattle for British post-medieval sites (with specific reference to unimproved British longhorn cattle) in B. Wilson, C. Grigson and S. Payne (eds) Ageing and Sexing Animal Bones from Archaeological Sites Oxford: British Archaeological reports (British Series 109), 37-54.

ARMITAGE, P. L. and CLUTTON-BROCK, J. (1976) A system for classification and description of the horncores of cattle from archaeological sites. Jour. Archaeol. Sci. 3, 329-348.

ARMITAGE, P. L., LOCKER, A. and STRAKER, V. (1987) Environmental archaeology in London: a review in H.C.M. Keeley (ed) op cit. 252-331.

BAILEY, G. and GRIGSON, C. (1987) Vertebrate archaeozoology in P. Mellars (ed) Research Priorities in Archaeological Science London: Council for British Archaeology, 17-22.

BELL, M. (1984) Environmental archaeology in South West England in H.C.M. Keeley (ed) op cit., 43-133.

COY, J. and MALTBY, M. (1987) Archaeozoology in Wessex in H.C.M. Keeley (ed) op cit., 204-251.

GERRARD, C. (in press) A regional approach to faunal data from Somerset in N.D. Balaam, B. Levitan and V Straker (eds) *Studies in Palaeoenvironment and Economy in South West England* Oxford: British Archaeological reports (British Series 181), 81-88

GRANT, A. (1975) The animal bones in B. Cunliffe *Excavations at Portchester Castle I: Roman* London: Society for Antiquaries (Research Report 32), 378-406.

GRIGSON, C. (1982) Sex and age determination of some bones and teeth of domestic cattle: a review of the literature in B. Wilson, C. Grigson and S. Payne (eds) *Ageing and Sexing Animal Bones from Archaeological Sites* Oxford: British Archaeological Reports (British Series 109), 7-24.

HENDERSON, C. G. (1985) Archaeological investigations at Alphington Street, St. Thomas in C.G. Henderson (ed) *Exeter Archaeology 1984/5* Exeter: City Council (Exeter Museums Archaeological Field Unit), 1-14.

KEELEY, H. C. M. (ed) (1984) *Environmental Archaeology: a Regional Review* London: Department of the Environment (Directorate of Ancient Monuments and Historic Buildings Occasional Paper 6).

KEELEY, H. C. M. (ed) (1987) *Environmental Archaeology: a Regional Review Vol II* London: Historic Buildings and Monuments Commission for England (Occasional Paper 1).

KENWARD, H. K., HALL, A. R., JONES, A. K. G. and O'CONNOR, T. P. (1984) Environmental archaeology at York in retrospect and prospect in H.C.M. Keeley (ed) *op cit.*, 152-179.

LEVITAN, B. (1985) Early 18th century horncores from Shooting Marsh Stile in C.G. Henderson (ed) *Exeter Archaeology 1984/5* Exeter: City Council (Exeter Museums Archaeological Field UNit), 15-18.

LEVITAN, B. (1987) Medieval animal husbandry in South West England: a selective review and suggested approach in N.D. Balaam, B. Levitan and V. Straker, (eds), *Studies in Palaeoenvironment and Economy in South West England* Oxford: British Archaeological Reports (British Series 181), 51-80.

LEVITAN, B. (forthcoming a) A method for calculating anatomical representation.

LEVITAN, B. (forthcoming b) The vertebrate remains from West Hill, Uley.

MACGREGOR, A. (1985) *Bone, Antler, Ivory and Horn. The technology of Skeletal Materials since the Roman Period* London: Croom Helm.

MALTBY, M. (1979) *Faunal Studies on Urban Sites. The Animal Bones from Exeter 1971-1975* Sheffield: University of Sheffield Department of Prehistory and Archaeology (Exeter Archaeological Report 2).

NODDLE, B. (1985) The animal bones in R. Shoesmith, *Hereford City Excavations Volume 3: the Finds* London: Council for British Archaeology (Research Report 56), 84-94 and microfiche M8.C8-M9.A13.

O'CONNOR, T. P. (1982) Animal bones From Flaxengate, Lincoln c 870-1500 London: Council for British Archaeology (The Archaeology of Lincoln Volume XVIII-1).

O'CONNOR, T. P. (1983) Aspects of site environment and economy at Caerleon fortress baths, Gwent in B. Proudfoot (ed) Site, Environment and Economy (Symposia of the Association for Environmental Archaeology 3) Oxford: British Archaeological Reports (International Series 173), 105-113.

O'CONNOR, T. P. (1984) Selected Groups of Bones from Skeldergate and Walmgate London: Council for British Archaeology (The Archaeology of York. The Animal Bones 15/1).

O'CONNOR, T. P. (1986) The animal bones in J.D. Zienkiewicz, The Legionary Fortress Baths at Caerleon II: the Finds Cardiff: National Museum of Wales, 224-248.

SCHMID, E. (1968) Beindrechsler, Hornshnitzer und Leimsieder im römischen Augst in E. Schmid, L. Berger and P. Burgin (eds) Provincialia. Fetschrift für Rudolf Laur-Belart Basel: Schwabe, 185-197.

WATSON, J. P. N. (1979) The estimation of the relative frequencies of mammalian species: Khirokitia 1972 Jour. Archaeol. Sci. 6, 127-137.

WILSON, B. (1978) Sampling bone densities of Mingies Ditch in J.F. Cherry, C. Gamble and S. Shennan (eds) Sampling in Contemporary British Archaeology Oxford: British Archaeological Reports (British Series 50), 35-361.

WILSON, B. (1985) Degraded bones, feature type and spatial patterning on an Iron Age site in Oxfordshire, England in N.R.J. Fieller, D.D. Gilbertson and N.G.A. Ralph Palaeobiological Investigations. Research Design, Methods and Data Analysis (Symposia of the Association for Environmental Archaeology 5B) Oxford: British Archaeological Reports (International Series 266), 81-93.

WILSON, B. and LEVITAN B. (forthcoming) The vertebrate remains from Claydon Pike, Gloucestershire.

Table 1. Cattle anatomical counts, Exe Bridge, Exeter

ELEMENT	EF	13th century N	13th century N/EF	14th century N	14th century N/EF	15th century N	15th century N/EF	Post-medieval N	Post-medieval N/EF
horncore	2	450	225.0	24	12.0	19	9.5	42	21.0
upper teeth	12	10	.8	5	.4	8	.7	9	.8
lower teeth	18	16	.9	5	.3	4	.2	4	.2
mandible	2	3	1.5	6	3.0	2	1.0	2	1.0
cervical	7	29	4.1	9	1.3	7	1.0	19	2.7
thoracic	12	10	.8	3	.3	11	.9	7	.6
lumbar	7	14	2.0	5	.7	3	.4	7	1.0
caudal	16	39	2.4	4	.3	4	.3	0	.0
ribs	24	70	2.9	16	.7	17	.7	16	.7
scapula	2	12	6.0	2	1.0	1	.5	1	.5
humerus	6	11	1.8	8	1.3	4	.7	11	1.8
radius	6	13	2.2	2	.3	5	.8	6	1.0
ulna	2	14	7.0	6	3.0	6	3.0	6	3.0
carpals	12	3	.3	1	.1	12	1.0	1	.1
metacarpal	6	15	2.5	10	1.7	9	1.5	10	1.7
pelvis	6	2	.3	2	.3	0	.0	4	.7
femur	6	14	2.3	4	.7	3	.5	4	.7
patella	2	2	1.0	0	.0	3	1.5	1	.5
tibia	6	12	2.0	13	2.2	13	2.2	7	1.2
astragalus	2	14	7.0	2	1.0	5	2.5	13	6.5
calcaneum	2	26	13.0	4	2.0	5	2.5	6	3.0
tarsals	6	3	.5	1	.2	3	.5	0	.0
metatarsal	6	15	2.5	11	1.8	4	.7	11	1.8
phalanges	24	58	2.4	10	.4	7	.3	26	1.1
TOTAL		855		153		155		213	

EF = expected frequency in skeleton except long bones (2 proximal ends + 2 diaphyses + 2 distal); ulna (proximal only); pelvis (2 ilium + 2 ischium + 2 pubis); ribs (heads).
N = fragment counts exclude those less than 25% complete (except epiphyses).

Table 2. Sheep anatomical counts, Exe Bridge, Exeter

ELEMENT	EF	13th century N	13th century N/EF	14th century N	14th century N/EF	15th century N	15th century N/EF	Post-medieval N	Post-medieval N/EF
horncore	2	467	233.5	43	21.5	14	7.0	25	12.5
upper teeth	12	66	5.5	15	1.3	10	.8	5	.4
lower teeth	18	42	2.3	4	.2	8	.4	8	.4
mandible	2	149	74.5	20	10.0	13	6.5	19	9.5
cervical	7	4	.6	0	.0	21	3.0	13	1.9
thoracic	12	1	.1	0	.0	2	.2	15	1.3
lumbar	7	5	.7	1	.1	13	1.9	6	.9
caudal	16	3	.2	1	.1	2	.1	3	.2
ribs	24	58	2.4	21	.9	24	1.0	41	1.7
scapula	2	26	13.0	3	1.5	7	3.5	8	4.0
humerus	6	25	4.2	11	1.8	14	2.3	13	2.2
radius	6	68	11.3	15	2.5	18	3.0	12	2.0
ulna	2	9	4.5	5	2.5	3	1.5	6	3.0
carpals	12	0	.0	0	.0	0	.0	0	.0
metacarpal	6	160	26.7	22	3.7	38	6.3	31	5.2
pelvis	6	27	4.5	1	.2	6	1.0	18	3.0
femur	6	31	5.2	12	2.0	14	2.3	15	2.5
patella	2	0	.0	0	.0	0	.0	0	.0
tibia	6	75	12.5	18	3.0	30	5.0	18	3.0
astragalus	2	4	2.0	0	.0	1	.5	1	.5
calcaneum	2	5	2.5	0	.0	2	1.0	1	.5
tarsals	6	0	.0	0	.0	0	.0	0	.0
metatarsal	6	206	34.3	33	5.5	31	5.2	27	4.5
phalanges	24	67	2.8	14	.6	9	.4	16	.7
TOTAL		1498		239		280		301	

Key: see Table 1

Table 3. Cattle horncore ageing results, Exe Bridge, Exeter

AGE CLASS	AGE RANGE	13th century N	%	14th century N	%	15th century N	%	Post-medieval N	%
infant	0>1yr	6	1.8	0	.0	0	.0	1	3.6
juvenile	1>2yr	87	26.4	4	25.0	1	5.9	3	10.7
sub-adult	2>3yr	34	10.3	1	6.3	1	5.9	2	7.1
young adult	3>7yr	127	38.5	6	37.5	5	29.4	10	35.7
adult	7>10yr	58	17.6	2	12.5	7	41.2	5	17.9
old adult	>10yr	18	5.5	3	18.8	3	17.6	7	25.0
TOTAL		330		16		17		28	

Age classes and ranges from Armitage (1982b, 42)

Table 4. Cattle epiphysial fusion results, Exe Bridge, Exeter

	13th century F	NF	14th century F	NF	15th century F	NF	Post-medieval F	NF
7-18 months								
scapula D	6	0	3	0	1	1	0	0
humerus D	2	4	13	0	3	1	2	5
radius P	11	1	21	0	5	5	4	3
phalanx 1 P	13	0	26	0	5	0	15	0
phalanx 2 P	2	0	9	0	2	0	4	2
% not fused		16		0		30		29
24-36 months								
metacarpal D	4	6	14	3	3	6	2	4
tibia D	3	2	11	6	8	4	0	4
metatarsal D	2	12	6	2	1	8	0	7
calcaneum P	4	3	12	4	3	4	2	4
% not fused		64		26		60		83
42-48 months								
humerus P	2	6	3	3	2	3	6	6
radius D	4	0	10	1	1	6	3	4
ulna P	0	3	1	1	1	2	1	2
femur P	3	7	8	4	4	2	1	4
femur D	1	6	7	1	3	1	2	4
tibia P	1	4	6	7	1	7	5	7
% not fused		70		33		60		60

F = fused. NF = not fused. P = proximal. D = distal.
Fusion ages from Grigson (1982, 22)

Table 5. Cattle anatomical counts, St. Katherine's Priory, Exeter

ELEMENT	EF	13th century N	13th century N/EF	13-14 century N	13-14 century N/EF	14th century N	14th century N/EF	14-15 century N	14-15 century N/EF	15th century N	15th century N/EF	16th century N	16th century N/EF
teeth	30	4	.1	1	.0	9	.3	7	.2	15	.5	166	5.5
mandible	2	3	1.5	0	.0	1	.5	1	.5	0	.0	62	31.0
vertebrae	42	2	.0	1	.0	9	.2	2	.0	8	.2	169	4.0
ribs	24	6	.3	3	.1	5	.2	1	.0	7	.3	73	3.0
scapula	2	2	1.0	0	.0	4	2.0	3	1.5	0	.0	13	6.5
humerus	6	5	.8	1	.2	3	.5	2	.3	1	.2	47	7.8
radius	6	3	.5	1	.2	1	.2	7	1.2	2	.3	47	7.8
ulna	2	1	.5	0	.0	0	.0	4	2.0	1	.5	26	13.0
carpals	12	2	.2	0	.0	1	.1	0	.0	0	.0	21	1.8
metacarpal	6	5	.8	0	.0	1	.2	2	.3	1	.2	71	11.8
pelvis	6	4	.7	1	.2	2	.3	0	.0	0	.0	23	3.8
femur	6	6	1.0	2	.3	5	.8	2	.3	2	.3	19	3.2
tibia	6	2	.3	2	.3	2	.3	1	.2	0	.0	69	11.5
tarsal	10	5	.5	2	.2	4	.4	2	.2	6	.6	77	7.7
metatarsal	6	2	.3	0	.0	2	.3	2	.3	0	.0	70	11.7
phalanges	24	7	.3	3	.1	4	.2	4	.2	12	.5	59	2.5
TOTAL		59		17		53		40		55		1012	

Key: see Table 1

Table 6. Sheep anatomical counts, St. Katherine's Priory, Exeter

ELEMENT	EF	13th century N	N/EF	13-14 century N	N/EF	14th century N	N/EF	14-15 century N	N/EF	15th century N	N/EF	16th century N	N/EF
horncore	2	0	.0	0	.0	0	.0	0	.0	1	.5	7	3.5
teeth	30	2	.1	0	.0	0	.0	2	.1	4	.1	27	.9
mandible	2	4	2.0	2	1.0	3	1.5	1	.5	4	2.0	57	28.5
vertebrae	42	10	.2	3	.1	10	.2	8	.2	10	.2	82	2.0
ribs	24	28	1.2	5	.2	13	.5	10	.4	18	.8	93	3.9
scapula	2	9	4.5	5	2.5	4	2.0	1	.5	4	2.0	22	11.0
humerus	6	23	3.8	10	1.7	5	.8	2	.3	14	2.3	87	14.5
radius	6	31	5.2	24	4.0	9	1.5	3	.5	18	3.0	78	13.0
ulna	2	1	.5	0	.0	1	.5	2	1.0	2	1.0	17	8.5
carpals	12	0	.0	0	.0	0	.0	0	.0	0	.0	0	.0
metacarpal	6	4	.7	0	.0	0	.0	0	.0	8	1.3	6	1.0
pelvis	6	8	1.3	8	1.3	3	.5	0	.0	9	1.5	36	6.0
femur	6	16	2.7	6	1.0	8	1.3	2	.3	4	.7	22	3.7
tibia	6	32	5.3	16	2.7	24	4.0	10	1.7	26	4.3	86	14.3
tarsal	10	2	.2	2	.2	4	.4	1	.1	3	.3	21	2.1
metatarsal	6	1	.2	1	.2	1	.2	0	.0	5	.8	2	.3
phalanges	24	1	.0	0	.0	1	.0	1	.0	0	.0	7	.3
TOTAL		172		82		86		43		130		650	

Key: see Table 1

Table 7. Summary of species present 1540-c.1570
St. Nicholas Priory, Exeter

SPECIES	N	%
cattle	2256	60.2
sheep/goat	1182	31.5
pig	163	4.3
dog	84	2.2
rabbit	24	.6
horse	17	.5
red deer	10	.3
fallow deer	7	.2
roe deer	3	.1
cat	2	**
brown rat	2	**
TOTAL	3750	
domestic fowl	71	71.0
goose	17	17.0
little gull	3	3.0
herring gull	3	3.0
duck	2	2.0
woodcock	1	1.0
lesser black-back gull	1	1.0
starling	1	1.0
rook	1	1.0
TOTAL	100	
cod	6	35.3
ling	5	29.4
conger eel	3	17.6
hake	1	5.9
haddock	1	5.9
bass	1	5.9
TOTAL	17	
indet. mammal	995	93.1
indet. bird	16	1.5
indet. fish	58	5.4
TOTAL	1069	
human	3	**
TOTAL	4939	

** = less than 0.1%

Table 8. St. Nicholas Priory: anatomical groups

	All contexts					Context 16				Context 17				Context 17-7				Context 17-8		
	N	AI	<25%	R		N	AI	<25%	%	N	AI	<25%	%	N	AI	<25%	%	N	AI	<25%
Cattle																				
SKL-JAW	430	1.94	98	16		18	.06	88	.9	94	.00	100	.0	55	.32	95	2.5	23	.92	65
HC	9	.50	89	25		0	.00	0	.0	6	.00	100	.0	0	.00	0	.0	1	.00	100
CAR-TAR	92	11.30	13	4		11	1.23	9	18.1	23	3.12	17	28.1	19	2.21	26	17.4	5	.99	0
MC-MT	126	17.54	32	2		15	1.57	13	23.2	19	3.54	16	31.9	25	3.93	28	31.0	7	.45	14
FHAL	62	7.53	2	8		15	1.76	7	26.0	12	1.50	0	13.5	13	1.63	0	12.9	10	1.26	0
RIB	378	.36	92	26		45	.04	91	.6	63	.12	87	1.1	89	.00	100	.0	69	.14	30
VERT	432	2.54	62	13		73	.34	68	5.0	48	.14	87	1.3	99	.56	69	4.4	56	.40	59
SCAP-PEL	190	1.86	89	17		19	.29	5	4.3	22	.09	95	.8	52	.19	94	1.5	28	.46	79
HUM-RAD-ULN	239	7.91	62	7		29	.65	69	9.6	31	1.51	52	13.6	67	1.39	69	14.9	36	1.07	67
FEM-TIB	263	5.89	51	10		43	.84	81	12.4	29	1.08	83	9.7	73	1.94	77	15.3	32	.75	78
							6.78				11.10				12.67				6.47	
Sheep																				
SKL-JAW	135	2.85	86	11		15	.21	37	1.6	13	.00	100	.0	45	.00	100	.0	5	.42	63
HC	5	1.63	50	18		0	.00	0	.0	3	.00	100	.0	3	.13	0	.8	0	.00	0
CAR-TAR	31	6.30	0	9		9	1.87	0	14.3	5	1.03	0	11.9	4	.85	0	5.4	6	1.34	0
MC-MT	48	8.60	13	6		17	2.80	12	21.4	2	.50	0	5.8	7	1.15	14	7.3	6	1.25	17
FHAL	20	2.51	0	14		9	1.13	0	8.6	2	.26	0	3.0	2	.25	0	1.6	0	.00	0
RIB	225	1.17	61	21		39	.18	72	1.4	33	.09	82	1.0	42	.12	53	.8	28	.12	64
VERT	162	2.36	28	12		36	.57	22	4.4	32	.46	34	5.3	17	.41	12	2.6	18	.26	33
SCAP-PEL	155	9.40	45	5		26	1.61	42	12.3	30	1.02	57	11.7	37	2.85	43	16.2	23	.92	57
HUM-RAD-ULN	203	21.86	17	1		25	2.91	8	22.2	27	2.02	37	23.2	57	6.44	16	41.0	17	1.80	24
FEM-TIB	193	15.25	40	3		27	1.81	41	13.5	36	3.31	33	38.1	53	3.50	66	22.3	19	2.25	16
							13.09				8.69				15.70				8.36	
Pig																				
SKL	26	.19	92	27		2	.00	100	.0	0	.00	0	.0	3	.32	75	12.6	3	.00	100
CAR-TAR	6	1.21	17	20		3	.32	0	15.5	2	.04	50	10.3	1	.25	0	9.9	0	.00	0
MC-MT	7	.71	0	24		3	.19	0	9.2	1	.07	0	17.9	2	.10	0	4.0	0	.00	0
PHAL	7	.88	0	23		2	.26	0	12.6	0	.00	0	.0	1	.13	0	5.1	0	.00	0
RIB	18	.02	89	29		7	.01	14	.5	0	.00	0	.0	1	.00	100	.0	1	.00	100
VERT	20	.15	45	28		5	.03	60	1.5	0	.00	0	.0	5	.04	40	1.6	1	.02	0
SCAP-PEL	23	1.50	35	19		3	.32	33	15.5	0	.00	0	.0	5	.65	20	25.7	6	.34	33
HUM-RAD-ULN	25	2.15	20	15		6	.62	17	30.1	4	.22	50	56.4	10	.83	20	32.8	2	.25	0
FEM-TIB-FIB	21	.63	63	22		9	.31	45	15.0	1	.06	0	15.4	4	.21	50	8.3	0	.00	0
							2.06				.39				2.53				.61	

N - number of identified fragments   AI - anatomical index   <25% - proportion of identified fragments less than quarter complete   R - rank

SKL - skull; JAW - mandible; HC - horncore; CAR - carpals; TAR - tarsals; MC - metacarpal; MT - metatarsal; PHAL - phalanges; VERT - vertebrae; SCAP - scapula; PEL - pelvis; HUM - humerus; RAD - radius; ULN - ulna; FEM - femur; TIB - tibia; FIB - fibula.

DECIDING PRIORITES WITH URBAN BONES:
YORK AS A CASE STUDY

Terry O'Connor

It is the purpose of this paper to examine the way in which the study of an exceptionally large archive of bones from archaeological sites in one city has evolved and adapted in the face of changing research priorities and the practicalities of implementing them. To some extent, York is an unusual case, having seen an intensity of archaeological work matched in Britain only in London. However, the extreme problems of volume of material and potential repetition of research have forced a consideration of the underlying premises of archaeological bone studies, and some of the conclusions which have been reached, and the compromises which have been made, will be relevant beyond the confines of this one research programme. This paper does not lay down guidelines for others to follow, however. Different circumstances require different measures, and some colleagues may disagree strongly with some of the sampling strategies and corner-cutting exercises which are described below. The author is unrepentant: just as there is more than one way to skin a cat, so there are several ways of studying its skeleton.

To set the background in which this work has taken place, archaeological excavation in York is mainly undertaken by the York Archaeological Trust. Working in close liaison with Y.A.T., but separate from it, is the Environmental Archaeology Unit, a research team within the Biology Department of the University of York, largely funded by grants from the Historic Buildings and Monuments Commission for England (O'Connor et al 1984). Bones are studied as one element in a range of plant and animal remains from archaeological deposits in the city. In parts of York, the accumulation of occupation debris amounts to seven or eight metres in depth, much of which is waterlogged. Preservation of all types of biological remains is often excellent and copious, and fragments of bone are no exception. The integration of bone studies with a range of other biological remains has been an important influence in the development of techniques and approaches. If bones are the only biological remains uncovered from a site, it is easy to overstate their value, and to regard the cataloguing of each specimen as an end in itself, rather than as a means to an end. When the bones are just one class of fossil recovered from the same deposits, sometimes even from the same soil samples, as beetles, seeds, bits of timber, shells, fly puparia and so on, then this unfortunate tendency is more likely to be avoided.

Which bones should I record?

Faced with an archive of around a couple of million bones, and with more coming out of the ground daily, the "catalogue everything" approach clearly had to be abandoned at the outset. The backlog currently in hand would require something of the order of twenty years of dogged cataloguing and tabulation, the eventual product being a huge volume of data which answered some questions many times over whilst leaving others unanswered.

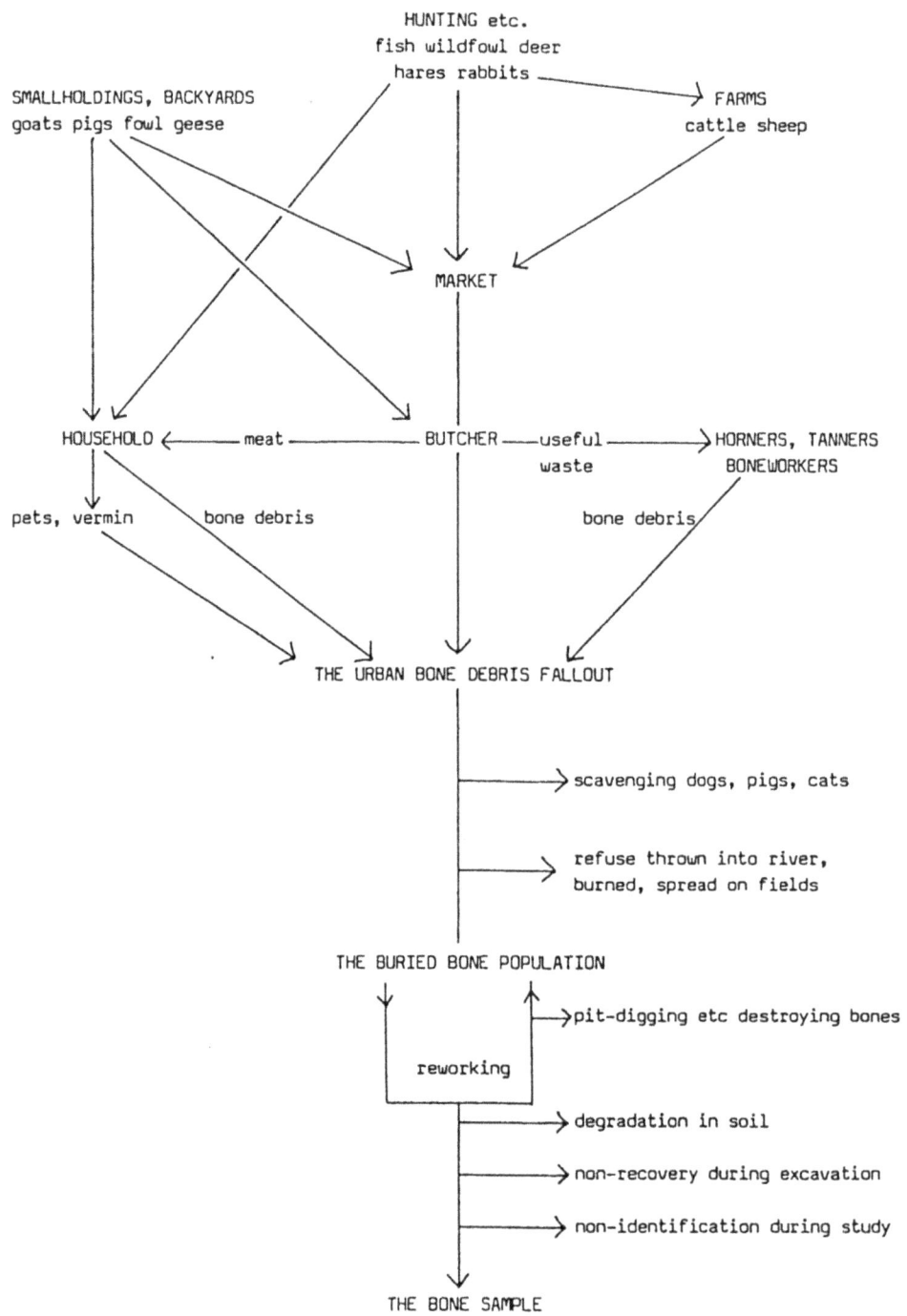

Figure 1. Simplified flow diagram for bones from livestock and hunted game to the archaeological bone sample.

The principal aim of our work has been to characterise bone deposition across the city through the 1900 or so years of its occupation in order to answer questions about human activity. The first priority has been to decide which bones from which deposits should be studied. No amount of care in devising recording or analytical methods can compensate for having selected the wrong sample in the first place.

An important aspect of this research aim has been the study of the husbandry and biology of the livestock exploited by the urban population, and of the pests and commensal vertebrates which exploited them. These two research areas are placed in this order because we are dealing with a city, not with a site of livestock production. The excavated bone sample is many stages removed from the processes of trade, butchering and domestic consumption, each of which will have left its distinctive imprint upon the deposited assemblage, but still more stages removed from the husbandry regime which originally produced the livestock. Figure 1 attempts to map a greatly simplified flow diagram for bones from livestock to the urban archaeological sample. Samples are selected for study on the basis of what they can reveal about human activity within the city because that is a more direct, or at least less indirect, line of enquiry. In fact the two research aims rarely conflict: samples which provide good evidence of human activity within the city usually also provide useful data relevant to livestock morphology and husbandry.

The first criterion on which selection is made is that preservation and recovery should be good. There is little point in basing a detailed analysis of butchery procedures or age selection at slaughter on a small sample of badly-preserved bones which were excavated by pick-axe. Good or bad preservation is easy enough to assess, but quality of recovery can be a more contentious point. Studies of bones from urban archaeological sites have up to now been based largely on assemblages acquired by hand-collection, the picking-out of bone fragments as a deposit is trowelled away. Of course this introduces a bias: small bones are less likely to be recovered. Despite the weight of evidence to the effect that this bias is very substantial (Payne 1975; Clason & Prummel 1977; Levitan 1982), it is often ignored when the data are discussed (eg. see O'Connor 1982; Jones 1984). One of the major innovations in urban archaeology which has been made in York has been the routine wet-sieving of large soil samples of 50-200kg for the recovery of small bones (Kenward, Hall and Jones 1980). This at least allows the extent of the hand-collection bias to be assessed, and the availability of sieved bone samples for use as a sort of quality control has become an important criterion in the selection of hand-collected bones for study. Latterly, the process has been enlarged to include the sieving of substantial deposits, hundreds of kilograms per sample, on a 10mm mesh. Obviously numbers of small bones will be lost through a mesh of this size, but at least the bias is known and constant. The problem with hand-collected bones is that the efficiency of operator recovery is wildly and unpredictably variable.

Given satisfactory preservation and recovery, the next step is to refer to the archaeological records for the site in question to determine as far as possible the origin of the bones. Priority is given to bone groups from closely-dated deposits, or from features which can be directly related to a specific structure or activity, such as a group of pits associated with a single phase of occupation of a particular building. Much of the bone debris from a typical urban site consists of small numbers of fragments from deposits representing the gradual build-up of ground surfaces, or from demolition deposits, or, worst of all, from the fills of stone-robbing trenches. A useful concept here is that of "background". Human occupation sites can be seen as having an inevitable background fall-out of bone fragments into any accumulating deposit, such as house floors, fills of pits, post-holes and road surfaces. In many circumstances this bone debris will not represent nearby human activity, and will contain a sample of all the different activities in different places around the town. The typical content of background debris will vary a little from one century of deposition to the next, and its information content is negligible. Learning to recognise bone assemblages which are merely "background", so as to avoid making more than a minimal record of them, has been one of the most useful developments in assigning priorities to bone groups. Obviously this could only be achieved once sufficient material from the town had been recorded to allow familiar patterns of bone deposition to be recognised.

In the specific context of York's archaeology, priority is given to certain periods of history, in particular to the 4th to 9th centuries AD and to the post-medieval period, and to bone deposits clearly derived from specific industrial activities, stretching the term 'industrial' to include primary butchering and other activities connected with the processing of animal carcasses. These industrial assemblages give information which can be linked to a specific activity or suite of activities, and are thus relatively simple to interpret. The most substantial example published to date is the huge assemblage of sheep foot bones recovered from 17th-18th century deposits in Walmgate (O'Connor 1984), and clearly referable to the activities of a tawyer or fell-monger. High priority is also given to sieved bone samples which yield a large assemblage of bones of small vertebrates. Depending on the circumstances of deposition, these assemblages may be rich in bones of fish and household pests such as rats and mice, or they may give a high diversity of rodent, insectivore and amphibian taxa which indicate the surrounding environment at the time of deposition. In either case, the bones will give information useful in the reconstruction of the depositional and taphonomic history of the deposit from which they came, quite apart from their interpretation in terms of diet or living conditions.

During the initial selection of samples for study, close co-operation with the people responsible for other aspects of the post-excavation study is invaluable, in particular with pottery researchers. As bone fragments are not intrinsically date-attributable, unlike potsherds, residuality can be hard to assess. In an urban settlement, pressure on living space leads to extensive reworking of earlier occupation deposits, through the digging of pits and wells, and residuality of bone can be a major problem. If pottery studies have shown that, for example, 60% of the pottery from a medieval pit comprises abraded sherds of terra sigillata, then the bone fragments from the same deposits can hardly be relied upon to give a useful indication of medieval disposal practices.

The preliminary selection of samples for study may seen to be a long and complex business. In fact, a practical and academically useful selection is often made by means of a 'rummage survey'. This entails a few quiet hours spent among the stored boxes of bones, armed with notes describing the stratigraphy and dating of the site, lifting off lids and generally poking about amongst the bones. Preservation can be assessed, particularly distinctive groups (such as concentrations of horn cores or dead cats) can be located, and a clear impression of the potential of the assemblage can be gained. Having undertaken such a preliminary examination, one is in a position to recommend to the archaeologist what bones are worth further study.

How to Record

Having selected the bone groups which will be recorded, the next questions are: how will the record be made, and in what detail? There are basically two approaches to recording. All the data pertaining to a particular bone can be kept together in a single record, or the different categories of data can be recorded separately in a series of different records, each one devoted to a particular category. The single record approach has the advantage that all the information about one fragment is retained together, but has the disadvantage that the dataset which results from an assemblage of any size is so huge and complex that resort to computers becomes essential (Hardy 1982). The 'many records' approach has the advantage of comparative simplicity in data handling. If one wishes to examine the biometrical data from a particular assemblage, they are available in a separate file. The disadvantage of such an approach is that the data obtained from one specimen are split up at the point of recording, and cannot be reunited, short of individually numbering specimens.

A computer based, single-record system was used for some time at York, producing lengthy catalogues of what were, in effect, bone-by-bone descriptions. However, the need to use the computer even to scan through the results from a particular layer, coupled with the disastrous effects of hardware failure, led to a reappraisal. In practice, the unit of data which is used for analysis and interpretation is not the individual specimen, but a summary of data recovered from each sample. The sample, in this sense, could either be the bones recovered from a single soil sample, or the specimens hand-collected from a single excavated context. Accordingly, a different approach to recording was adopted, giving priority to the production of a collective description of the bones in each sample.

One of the main advantages of the change of approach is that the initial record is made on a paper pro forma, from which data can be abstracted for a computerised database, which is much more condensed and faster to manipulate than even the most efficient bone-by-bone catalogues. Use of the pro forma requires no competence with computers, merely with bones, and allows the number of staff working on bones to be increased when necessary without requiring either an expensive proliferation in hardware or the sort of period of training which short-term project-funding does not allow. In short, it makes the job more cost-effective.

Abandoning the catalogue approach to bone recording rather went against conventional wisdom, requiring, as it did, a complete redefinition of what one was trying to record. Once again, day-to-day contact with colleagues working on different groups of organisms was an important catalyst. Archaeobotanists do not find it necessary to make seed by seed lists of the contents of their samples, nor do entomologists catalogue all the attributes of each piece of beetle in theirs. In other areas of bioarchaeology, the priority has been to make a record of the sample, not of the specimen, and to a large extent, the change in recording methods which has been effected at York has merely brought bones into line with the many other categories of plant and animal fossil.

In how much detail should data be recorded?

When deciding the degree of detail in which data should be recorded, priority has been given to acquiring the dataset which experience has shown will be required for the analyses likely to be undertaken in order to answer archaeological and biological questions immediately in hand. The priority in selecting methodological procedures for taxon quantification, age estimation, biometry, and so on, has been to allow hypotheses to be tested, rather than to create an archive with a view to colleagues reworking the data at some time in the future. There are arguments for and against recording for posterity which are to some extent unresolved. Suffice it to say that in this instance the decision was taken to attend to current needs, rather than to compromise research in hand in the hope of making a record which might be of value in the future.

Quantification

Taxon quantification based on fragment counts is an approximate and unsatisfactory technique for assessing the relative proportions of taxa in the original killed population (Fieller and Turner 1982). The presence of whole or partial skeletons in the assemblage will create a serious bias, by contributing many bones from a single individual, and it might be argued that 'minimum number of individuals' methods would be more appropriate. However, it is highly debatable whether the concepts of individuals and killed population have any relevance when considering the bone debris from urban settlements. The killed population is many stages removed from the recovered assemblage: much more so than is the case with bone assemblages from hunter-gatherer or subsistence agriculture sites. What is the killed population from which the sheep bones in a 13th century pit in York were derived? The individual sheep involved are likely to have originated in several different flocks, and some may have been driven to the city from tens of kilometres away. Calculation of minimum numbers of individuals is seldom useful with bones from urban excavations as the unit of distribution of domestic livestock within the city is likely to have been the joint, or the side of bacon, rather than the whole cow or pig, a point which has been well made by Armitage (1982, 95). Recovering an estimate of three cattle and fifteen pigs from one particular pit could indicate that the deposited assemblage originally comprised three whole cattle carcases and fifteen entire pigs, or it might show that a particular household bought in three smoked shoulders of beef and fifteen pigs' heads for brawn. Each reconstruction would prompt a quite different interpretation. Having rejected quantification methods based on numbers of individuals, one is left with fragment counts as a crude means not of reconstructing farmed or slaughtered populations, but of describing what was actually found in the recovered assemblage. Providing interpretation is kept at that level, fragment counts have a use in urban bone studies.

Given that farm livestock circulated within towns as butchered joints of meat and ancillary meat products, the quantification of skeletal elements must be given a high priority. A variety of different ways of counting skeletal elements have been proposed, most of them based on variants of Watson's (1979) concept of 'diagnostic zones'. Small, structurally simple, bones such as astragali can be quantified fairly easily. It is a comparatively simple matter to take whole and fragmentary cattle astragali and to decide on the smallest number which must be present. Limb bones and skulls present more complex difficulties, and the problem of what to do with identifiable diaphysial fragments, in particular, remains unresolved. One compromise is to count only those limb bone fragments which have all or part of an epiphysis appending. Though procedurally convenient when dealing with small assemblages, it is felt that this can lead to an unacceptable loss of information when making a detailed study of large assemblages. For one thing, some epiphyses are particularly susceptible to destruction. A good example is the proximal epiphysis of the cattle humerus, which is rarely encountered in post-Roman assemblages from York, though the shaft, heavily gnawed at the proximal metaphysis, is quite common. Furthermore, the diaphysis itself may be an important zone of carcass division. Roman deposits at 24-30 Tanner Row, York, provided a good example. Several contexts of late 2nd century date gave assemblages which consisted almost exclusively of fragments of heavily butchered cattle femur, tibia, humerus and radius diaphysis. A quantification based on epiphyses would have ignored over 90% of the identifiable bones. The system which has been adopted for large assemblages, therefore, gives equal weight to diaphysial and epiphysial diagnostic elements by defining zones to which shaft fragments can be allocated. This is not the place to go into the procedure in great detail. This allows detailed recording of fragmented material and a close degree of homology between taxa. The latter point is an important one, as it allows a further means of taxon quantification through an element by element comparison.

If the concept of the urban bone assemblage as a sample of butchery and food debris is taken to its logical conclusion, then it can be argued that simple taxon quantification, in the form of a comparison of the relative abundance of the bones of taxon A as against taxon B, is less useful as an interpretative tool than a quantification based on an *a priori* zoning of the carcasses of food animals. On this model, the assessment of an imaginary assemblage would change from the familiar 'mostly cattle, fewer sheep, still less pigs and hardly any poultry' to mostly cattle vertebrae and hindquarters, sheep forequarters, and pig trotters; with rather less sheep and pig vertebrae and ribs; and a background scatter of other bits.' In effect, different part of the carcass of the main food animals come to be treated as taxa in themselves. Apart from making more sense in the context of an economic interpretation, such a procedure gives a dataset with many more 'taxa' and greater differences between samples. In practice, such data are more susceptible to objective statistical measurement of diversity and between-sample distances. Rank abundance analyses can be undertaken, treating the bone assemblage in a way analogous to the analysis of multi-species populations, and such analyses often result in a clear characterisation of what is particular to the sample and what is merely background.

An important tool in the investigation of husbandry practises is the reconstruction of age at death. Again, the complexity of urban markets puts several additional layers between the living herd or flock and the recovered assemblage. The main aim of age at death reconstruction in York has been to describe the age structure of the livestock slaughtered in the city at different periods. Attempts to extend this interpretation to the construction of herd age structures and survivorship curves are bedevilled by our lack of knowledge of the marketing process which brought livestock into Roman and medieval towns. The notion that urban butchers dealt directly and individually with the farmers who raised the livestock is probably simplistic. As soon as we have reliable literary records, in the late medieval period, it is clear that we are dealing with a remarkably complex sequence of middlemen and graziers, placing several layers of age-group selection and life history between the producer and the butcher. Herd population structure and life expectancy are useful concepts in the archaeology of subsistence communities, but it is highly debatable whether they have any place in the study of urban assemblages. Until the sources of supply of towns in the past are better understood, and the relationship of the excavated assemblage to the original herd or flock made clear, age at death reconstruction in urban assemblages is best limited to considering what age groups of animals were selected for slaughter at different times and places, and what this might show about the economic pressures within the town.

Obviously there are several routes to an age at death reconstruction, and work in York has been based on mandibular eruption and attrition. This is not to decry the use of epiphysial fusion as a technique for age estimation, but there are interpretational problems with this technique (Watson 1978), and the fact that analysis of epiphysial fusion is based on elements from throughout the skeleton can be problematical in the urban context, where livestock would have been transported as joints rather than individuals. If the preferred pattern of butchery of, for example, pigs changed according to the age of the individual carcass, and if distribution of different cuts of pork within the town was not spatially homogenous, then a simple interpretation of epiphysial fusion data from one series of deposits could be very misleading. For these reasons, age estimation in York bones has concentrated on mandibular eruption and attrition, using the wear stages defined by Grant (1982, 92-4) as the basis of the record. However, Grant's interpretational technique, based on summed 'mandibular wear scores', is not used. Epiphysial fusion data, where available, are used as a cross-check on the mandibular data, and only in circumstances where differential disposal of immature and adult carcasses is not suspected. In presenting the data, the emphasis has been on defining developmental age groupings rather than calendrical ones. This allows a greater degree of comparability with the data obtained by other workers, and sidesteps the problem of not actually knowing at what age the lower third molar erupted in a 15th century sheep. Furthermore, it seems unlikely that the farmers of Roman to medieval Yorkshire knew the age of their livestock in terms of months or years. Then, as to a large extent now, it was the stage of development reached by the animal which determined whether or not it was slaughtered, mated, sold to a neighbour, or whatever. A comparison of developmental stage at death is thus probably more directly relevant than an attempt to compare absolute age distributions. Only when there is a need to time the slaughter of livestock to a particular month or season of the year does the need to convert stage of development to absolute age actually arise, and then the assumptions which must be made render the whole procedure very approximate and tenuous (eg. see Lauwerier 1983).

One line of research which deals with the data in wider groupings than the individual context or sample is the investigation of biometry. Once again, the emphasis has been on answering specific questions, rather than archiving data for their own sake. The main subject to have been tackled is the simple question of the extent to which the gross size and skeletal morphology of the main domesticates varied from period to period. In order to explore this topic, it has been necessary to collect data which describe the size and conformation of some of the main elements of the skeleton, and which are commonly recovered in measurable condition. This last point cannot be over-stressed. There is little point in attempting to base a biometrical study on measurements which are rarely available. The standard set of descriptions of bone measurements is used (von den Driesch 1976), though the full set of variates described by von den Driesch has steadily been whittled down with experience to a short-list of those which are at least occasionally available. The routine recovery of the bones of rodents and amphibians from sieved samples has prompted some investigation of the potential of biometrical work on taxa other than the familiar domestic livestock and pets. To date, this work has either been taxonomic, differentiating Mus and Apodemus specimens, or demographic, investigating age-related size classes in frog bones (O'Connor 1988).

Several other areas of bone studies have been given high priority. The recording of specimens which exhibit abnormalities attributable to disease or trauma has been one such, the emphasis being on describing the lesions in consistent terms so as to facilitate comparisons and objective diagnosis. Certain pathological conditions have become sufficiently familiar to justify attempts at quantification. Some manifestations of arthropathy in cattle, in particular osteoarthritis in the metapodio-phalangeal joints and acetabulum, occur at frequencies of 1-2% of all acetabulae or metapodia. These arthropathies, amongst others, are probably largely attributable to the uses to which cattle were formerly put, in particular hauling carts and ploughs. These occupational disorders are obviously of particular archaeological significance. At the other extreme, single instances of a range of pathological abnormalities have been encountered. Whilst intrinsically interesting, and worth noting so as to set modern veterinary records in an historical context, these specimens add little to the archaeological interpretation of the assemblage.

Non-metrical skeletal traits are potentially valuable in characterising the different gene pools being sampled by an urban market. Although some work has been undertaken in this field (eg. see Noddle 1978; Andrews and Noddle 1975), there has been little systematic use of the results as part of the archaeological process. The procedure at York has been to make a systematic record of the occurrence of two traits in cattle and sheep mandibles, namely the congenital absence of the second premolar and reduction of the distal column of the third molar. These have been chosen as being reasonably common, easy to score whilst recording eruption and attrition data, and relatively unambiguous. The important point to note is that normal specimens are recorded as well as abnormal ones. Here and there in the literature one encounters reports which describe an interesting specimen of absence of second premolar in a cattle mandible without giving any indication of whether this one specimen represented 10% of all cattle mandibles in that sample or only 0.1%. By concentrating on two fairly common traits, sufficient data have been amassed to give an impression of the 'usual' frequency of these traits in cattle and sheep mandibles, making it possible to draw attention to samples in which the frequency is either strikingly high or low, with possible implications about gene pool size and inbreeding. Although still in its infancy, this investigation seems particularly useful in the urban context, as it offers some hope of disentangling the many different sources of supply which towns must have had. Somewhat ironically, little attempt has been made to apply the study of non-metrical traits to rodent and insectivore bones from urban sites, despite the wealth of modern studies on epigenetic polymorphism in rodents. Although plenty of theories about the status of ship rat (Rattus rattus) populations in early English towns have been advanced (Armitage et al 1984; Davis 1986), this obvious means of investigating the rival claims of population continuity or repeated introduction has yet to be applied.

Conclusions

To summarise the points made in this paper, the key to the cost-effective study of bones from urban archaeological deposits is careful selection. An urban excavation will produce samples of bones which may yield useful information about human activities in and around the town, and others which are merely 'background', and of little value. Residuality can also be a serious problem, and close liaison with pottery researchers and others studying different materials from the same deposits is essential. Different periods of a town's development may merit particular study. In the case of York, any bone samples from the 4th to 9th centuries, or 16th to 18th centuries are given a high priority. Deposits of bones derived from specific craft or industrial activities may also be particularly informative as the papers of MacGregor and Serjeantson in this volume show.

Sieving is especially important as a control on the level of bone recovery. Even a coarse mesh, such as 10mm, will give better recovery than collection by hand during excavation. The important point about bone recovery by sieving is that the level of recovery is constant. Sieving also tends to direct recording procedures towards producing a record of the contents of a sample, rather than cataloguing individual specimens. This tends to make the bone data more compatible with studies of other biological remains from the same deposits.

Analytical techniques must take account of the complex route between the original herds and the urban deposit. Some species quantification methods can be ruled out as inappropriate. Reconstruction of herd age structure is problematical when the town may have drawn on many different sources of livestock, and this also complicates the interpretation of biometrical variation. Recording the distribution of different carcass elements of the main domesticates is important, as the unit of distribution within a town was probably the joint of meat, rather than the individual animal.

The main principle underlying the recording and investigation of bone samples from archaeological deposits in York has been the importance of attending to the job in hand, of answering current research aims rather than attempting to predict and satisfy the needs of future researchers by bequeathing them a detailed archive. This has been in part a deliberate intent and in part a pragmatic response to the need to extract information quickly and efficiently from what could easily become an overwhelming quantity of raw material. If the bones in question were subsequently going to be destroyed or reburied, then this selective information-grabbing would be difficult to justify. However, the primary archive of bones will be retained and curated by the designated receiving museum, the Yorkshire Museum, and will presumably be available for future research. In the circumstances, then, it has been justifiable to ascribe priorities on the basis of present needs.

There is another argument in favour of this rather selective approach. Whereas one large excavation might adequately represent the bone debris deposited at a small rural site, as at Danebury (Grant 1985), the complex pattern of deposition in a town will only gradually become apparent as more and more excavations are made into deposits of different dates. Until such time as bone samples are available from all periods of York's history, and from a range of different types of occupation within each period, the primary need will be to collect information in quantity, but at a low level of detail, in order that long-term trends and coarsely-defined patterns of activity and deposition may be seen. Only when this framework of knowledge is relatively complete will it become possible to determine the areas in which very detailed work is required to answer specific, narrowly-defined questions. For the present, the priorities are to select the best assemblages available in terms of preservation, recovery, and relevance to the archaeological record, and to concentrate on accumulating information based on secure (or, more honestly, less insecure) methodological procedures.

REFERENCES

ANDREWS A.H. and NODDLE B.A. (1975) Absence of premolar teeth from ruminant mandibles found at archaeological sites. Journal of Archaeological Science 2. 137-44.

ARMITAGE P.L. (1982) Studies on the remains of domestic livestock from Roman, medieval and early modern London: objectives and methods. In Hall, A.R. and Kenward, H.K. (eds) Environmental Archaeology in the Urban Context. London: C.B.A. Research Report 43. 94-106.

ARMITAGE P.L., WEST B. and STEEDMAN K. (1984) New evidence of black rat in Roman London. The London Archaeologist 4(14). 375-383.

CLASON A.T. and PRUMMEL W. (1977) Collecting, sieving and archaeological research. Journal of Archaeological Science 4. 171-5.

DAVIS D.E. (1986) The scarcity of rats and the Black Death: an ecological history. Journal of Interdisciplinary History 16(3), 455-70.

VON DEN DRIESCH A. (1976) A Guide to the Measurement of Animal Bones from Archaeological Sites. Peabody Museum Bulletin 1. Harvard: Peabody Museum.

FIELLER N.R.J. and TURNER A. (1982) Number estimation in vertebrate samples. Journal of Archaeological Science 9(1), 49-62.

GRANT A. (1982) The use of tooth wear as a guide to the age of domestic ungulates. In Wilson, B., Grigson, C. and Payne, S. (eds) Ageing and Sexing Animal Bones from Archaeological Sites. Oxford: British Archaeological Reports British Series 109. 91-107.

GRANT A. (1985) Animal husbandry. In Cunliffe, B. Excavations at Danebury. Vol II The Finds. London: Society of Antiquaries. 496-548.

HARDY E.A. (1982) A microcomputer-based system of recording bones from archaeological sites. In Aspinall, A. and Warren, S.E. (eds) Proceedings of the Micro-computer Jamboree. University of Bradford, 3rd April 1982. Bradford: University of Bradford. 11-17.

JONES G. (1984) Animal bones. In Rogerson, A. and Dallas, C. Excavations in Thetford 1948-59 and 1973-80. Norwich: East Anglian Archaeology Report 22. 187-92.

KENWARD H.K., HALL, A.R. and JONES A.K.G. (1980) A tested set of techniques for the extraction of plant and animal macrofossils from waterlogged archaeological deposits. Science and Archaeology 22, 3-15.

LAUWERIER, R.C.G.M. (1983) Pigs, piglets and determining the season of slaughtering. Journal of Archaeological Science 10, 483-88.

LEVITAN, B. (1982) The faunal remains. In Leach, P. Ilchester vol. 1: Excavations 1974-75. Western Archaeological Trust Excavation Monograph 3, 269-85.

NODDLE, B.A. (1978) Some minor skeletal differences in sheep. In Brothwell, D., Thomas, K.D. and Clutton-Brock, J. (eds) Research Problems in Zooarchaeology. London: Institute of Archaeology Occasional Publications 3, 133-41.

O'CONNOR, T.P. (1982) Animal bones from Flaxengate, Lincoln c. 870-1500. Archaeology of Lincoln 18/1. London: Council for British Archaeology.

O'CONNOR, T.P. (1984) Selected groups of animal bones from Skeldergate and Walmgate. Archaeology of York 15/1. London: Council for British Archaeology.

O'CONNOR, T.P. (1988) Bones from the General Accident site. London: Council for British Archaeology Archaeology of York 15/2.

O'CONNOR, T.P., HALL, A.R., JONES, A.K.G. and KENWARD, H.K. (1984) Ten years of environmental archaeology at York. In Addyman, P.V. and Black, V.E. (eds) Archaeological Papers from York presented to M.W. Barley. York: York Archaeological Trust. 66-77.

PAYNE, S. (1975) Partial recovery and sampling bias. In Clason, A.T. (ed) Archaeozoological Studies. Amsterdam: Elsevier. 7-17.

WATSON, J.P.N. (1978) The interpretation of epiphysial fusion data. In Brothwell, D., Thomas, K.D. and Clutton-Brock, J. (eds) Research Problems in Zooarchaeology. London: Institute of Archaeology Occasional Publications 3, 97-101.

WATSON, J.P.N. (1979) The estimation of the relative frequency of mammalian species: Khirokhitia 1972. Journal of Archaeological Science 6, 127-37.

GAZETEER OF SITES WITH ANIMAL BONES USED AS BUILDING MATERIAL

Philip L. Armitage

INDEX

1) ARCHITECTURAL FEATURES

    1.1 External decoration
    1.2 Interior decoration - "knuckle-bone floors"
    1.3 Floor decoration in "follies" - variation of the more traditional "knuckle-bone floors" found in private and public houses
    1.4 Patterned floors in summerhouses and other minor "follies"

2) CONSTRUCTIONAL WORKS - I: Industrial Contexts

    2.1 Horn core lined pits

3) CONSTRUCTIONAL WORKS - II: Domestic and Other Non-industrial Examples

    3.1 Packing/hardcore in house walls
    3.2 In repair works - where damage had occurred to the fabric of a building or part of a floor etc.
    3.3 Fastening slates to a church roof

4) CONSTRUCTIONAL WORKS III: Miscellaneous Categories

    4.1 Cesspits and soakaways
    4.2 Road foundations

5) AGRICULTURAL AND HORTICULTURAL USES

    5.1 Agricultural land drains
    5.2 Farmhouse roofs
    5.3 Walls built around meadows and market gardens

6) "ACOUSTIC RESONANCE VESSELS"

    6.1 Horse skulls placed beneath floors

1) ARCHITECTURAL FEATURES

1.1 **External decoration**

Site or place name: "Knuckle-bone house" (also known locally as "whale bone stores"), village of Cley on the north Norfolk coast.

Description: Cornice under the eaves of the roof is decorated with various horse, cattle and sheep bones (including first phalanges, metapodial bones, molar teeth and vertebrae) arranged in courses, tightly packed together with the articular ends of the limb bones facing outwards. In addition to the bones embedded in the cornice, there are square and rectangular panels set in the flint walls, between the windows, whose borders are formed by the distal articular ends of horse metapodial bones.

The house was built in the Dutch style for a merchant, in c. 1630; but it is not known whether the bone decoration was a contemporary or later feature. Whatever the original date of the design, it seems probable that many of the bone elements now embedded in the walls are recent replacements as it is unlikely that the originals could have survived so long in such an exposed situation without suffering from the effects of weathering.

The original source(s) of the bones is not known; but local oral tradition holds that they came from a "boneyard attached to a glue factory" that once operated in nearby Blakeney (dates unknown).

Present status: A well-known local landmark; now a book shop owned by Mr and Mrs P. Dearden.

Sources of information: Unpublished; pictures and description of the house supplied by Jennie Coy, Faunal Remains Project, Southampton University (1982, pers. comm.); historical details obtained from Mr and Mrs Dearden (1982, pers. comm.), B. Bland and Mrs. P. Suckling, both of Cley (1982, pers. comm.) and Sue Marjeson, Norwich Museum (1982, pers. comm.).

1.2 **Interior decoration** - "knuckle-bone floors"

Site or place name: Public house called "Antiquity Hall" (also known as "The Hole in the Wall" or by its sign of "Whittington and his Cat"). Formerly located in Oxford, on the south side of the street leading frm Hithe-Bridge to Rewley.

Description: Skelton (1823) described the floor as a "...pavement .. made of 'sheeps trotters' neatly composed in various compartments". Chambers (1863) also records tha the floor "...was paved with the bones of sheeps' trotters, curiously disposed in compartments".

From the published drawings by G. Virtue (reproduced in Skelton, 1823 plate 126) it would appear that the floor also incorporated cattle as well as sheep metapodial bones in the design; but this is nowhere mentioned in either Skelton (1823) or Chambers (1863). Inspection of the criss-cross pattern shown in Virtue's drawings suggests affinities with other contemporary bone floors elsewhere in Oxford (see examples from Holywell Street and Broad Street).

Exact date is unknown, probably late 17th century; the house is shown on Loggan's 1675 map of Oxford.

Present status: No longer extant; house survived until 1843 when it was demolished (see Hurst, 1894: 11).

Sources of information: Published accounts by Skelton (1823: 126), Chambers (1863: 765), Hurst (1894: 11) and Squiris (1928: 158). Additional historical information supplied by Mr. Graham, Central Library, Oxfordshire County Council.

Site or place name: Private house, No. 19 Holywell Street, Oxford.

Description: According to Hurst (1894) the floor was in the front room of the house and covered 255 sq. feet (71 sq.m.), and contained an estimated 24,460 bones. These bones were "halved and embedded, broad end upwards, in fine gravel, and when fixed in position, a mixture of thin lime and finer gravel appears to have been floated over the whole to give solidity and keep the surface light in colour. Whether hob-nails and street dirt, or rough sand and rubbing-stone were used to smooth the surface could not be decided. In one place, left of the hearth, the projections of the bones were left untouched...The borders, letters and figures of the design were of calves' bones, or those of smaller oxen; the dividing lines were smaller, and the filling in was of mutton bones or of lamb, and a few deer...In the initials the R is most probably the surname of the occupier, and W and E the Christian names of the husband and wife. The dates, 1701 and 1702, are easily recognised".

The source of the bones is not specified in Hurst's account, but may have been the slaughterhouse located behind the house.

Present status: No longer extant; was probably destroyed shortly after its discovery by workmen beneath a later floor of oak boards, during refurbishment of the house in 1893.

Sources of information: Published account by Hurst (1894: 10-11).

Site or place name: Private house formerly in Broad Street, Oxford.

Description: The floor "...was laid with 'trotter bones' in a pattern of squares, arranged angle-wise within a border. The pattern was defined by bones about 2in. (50 mm) square, rubbed or sawn to an even surface, and filled in with the small bones of sheep's legs, the knuckles uppermost, closely packed and driven into the ground to a depth of from 3in. (76 mm) to 4 in. (102 mm)". (Innocent, 1916: 159). Although not stated by Innocent, it would seem likely that the larger bones in the floor were from cattle.

Present status: No longer extant; destroyed during demolition of the house in 1869.

Sources of information: Innocent (1916: 159) who quotes an article that originally appeared in Building News for September 3rd, 1869.

Site or place name: Various properties located within the city of Oxford:

    i) Four private houses (including one known locally as the "Court House"), formerly located near the Hamel, St. Thomas' Parish.
    ii) Private house in Logic Lane.
    iii) Private house in George Street.
    iv) Original Ashmolean Museum building, Broad Street.

Description: Not described in the literature; all of the above are simply listed as examples of "knuckle-bone floors" and their dates of construction are also not indicated. The floors in the four houses in the Hamel as well as the one in the Ashmolean Museum were formerly well known local architectural curiousities. Workmen accidentally discovered the floor in George Street while demolishing the house. Contractors were also responsible for discovering the floor to the house in Logic Lane when they were digging up the ground in preparation for erecting scaffolding.

Present status: None of the above survived to the present day. The building known as the "Court House" stood at the south end of the Hamel until about 1880, when it was demolished (Squiris, 1928). The floors to the other houses were also destroyed by building activity at about the same time (see Hurst, 1894); the bone floor at the Ashmolean Museum was replaced by one of stone in the 19th century (Hurst, 1894).

Sources of information: Published accounts by Hurst (1894) and Squiris (1928). Additional historical information relating to the Ashmolean Museum was supplied by A. MacGregor (1982, pers. comm.).

Site or place name: 65 St. Aldates, Oxford.

Description: During excavations at the site, B. Durham of the Oxford Archaeology Unit discovered a small section of early 18th century flooring with five cattle metapodial bones forming a decorative line of infilling between the stone slabs. B. Wilson identified the bone elements as three proximal metacarpals, one proximal and one distal metatarsal, all of which had worn articular surfaces "presumably from the passage of footwear" (Wilson, 1984).

Present status: All traces of this feature were removed during the course of excavation.

Sources of information: Wilson in Durham (1984); archaeological details supplied by Durham (1984, pers. comm.).

Site or place name: "Bear and Ragged Staff" public house, Cumnor village near Oxford.

Description: No description available; history unknown.

Present status: Not known.

Sources of information: Mentioned in Hurst (1894: 12).

Site or place name: Private house, Fulbourn near Cambridge.

Description: "...the floor in that part of the house which was once an open

hall was thick with sheep bones...[forming] a sound durable [surface].."
(Evans, 1966: 200).

Present status: Not known.

Sources of information: Evans (1966).

Site or place name: Grammar school next to churchyard, Wantage, Berkshire.

Description: "Knuckle-bone floor"; date and details of construction not known.

Present status: No longer extant; destroyed about 1850.

Sources of information: Hurst (1894: 12).

Site or place name: Private house, Elstow near Bedford.

Description: No description available; history unknown.

Present status: Not known.

Sources of information: Hurst (1894: 12).

Site or place name: "New Inn" public house, 41-43 New Street, Salisbury.

Description: A section of a "knuckle-bone floor" was uncovered by workmen during alterations to the main bar room in 1972; presumed to be of 18th century date.

Proximal and distal articular ends of cattle metacarpal and metatarsal bones form the surface of the floor. All of these bones are broken in half and have their shafts driven into the earth to a depth of between 76 mm and 102 mm. Although only a rather small portion of the original complete floor survives there is sufficient remaining to show that the builders had made some attempt to produce a patterned effect: the distal metapodial bones are arranged in several neat rows, laid parallel to one another, and forming a border to a central square panel of proximal metacarpal bones.

Present status: The surviving section of this floor is on public display, protected by sheets of armoured glass, and lit for viewing. This now rare example of a "knuckle-bone floor" is in need or urgent attention; and unless measures are taken soon to improve the air flow under the glass viewing panels the rising moisture will over the next few years cause irreparable damage to the bones.

Sources of information: Unpublished; bones identified by P. Armitage during a visit to the public house in 1982. Additional information supplied by Mrs M. Spicer, proprietor of the "New Inn" and by Ms C. Conybeare, Salisbury and South Wiltshire Museum (1982, pers. comm.).

Site or place name: "King John's House", Romsey, Hampshire.

Description: An extensive area of "knuckle-bone floor" was uncovered during restoration work to the building in 1977. The floor is constructed from the distal ends of cattle metapodial bones, of which only the distal epiphysis and between one half and one third of the shaft remains. Marks left on the broken end of the shaft show where the bone was chopped in half by an axe or cleaver. Each bone was driven shaft end downwards into the natural brickearth, to a depth of 76 mm to 127 mm, depending on the length of the shaft, leaving the distal articular end uppermost. Although the underlying brickearth is relatively soft, no mortar was used to consolidate the structure, instead all the bones were packed closely together to produce a firm and stable surface.

Apart from the small area showing evidence of repair work, described below, the entire surface of the floor is covered by cattle metapodial bones arranged distal end uppermost. This uniformity in construction may be compared with the patterned "knuckle-bone floors" recorded from Oxford and Salisbury; the Romsey floor has the appearance of being functional rather than ornamental. The frequent movement of people in the room has resulted in considerable wear to the bones in the centre of the floor area, whilst those nearest the walls show little, if any, evidence of abrasion. In the most severely abraded bones, the distal condylar surface has been rubbed away leaving the underlying cancellous tissue exposed (Figure 1).

In one small area, the floor seems to have suffered damage and subsequently been repaired using an assortment of bones, including several complete sheep metapodial bones, three equid metatarsal bones (the small size of these indicating that they were either from donkeys or ponies) and various bones from joints of beef (eg. scapulae and tibiae). When first discovered the floor was presumed to be of medieval date; but inspection by P. Armitage revealed that it was probably much later than this - the large size of the cattle metapodial bones suggesting late 17th/early 18th century date. In order to test this assumption a sample of 48 metatarsal bones previously removed from the floor was compared with medieval, Tudor and early modern cattle metatarsal bones. From the size of the distal epiphysis it was clear that the Romsey bones were no earlier than the late 15th and early 16th century. Although the range of the distal widths in the Romsey sample compared favourably with the early Tudor cattle from London, their projected lengths greatly exceeded the largest of these 16th century beasts and resembled the metatarsal bones from the much taller 17th and 18th century cattle. Thus a date some time in the 17th or 18th century is suggested for the construction of the "knuckle-bone floor" in King John's House, based on the results of osteometric analyses. To obtain a more precise date, a sample of cattle metatarsal bones from the floor was submitted to the British Museum Research Laboratory for radiocarbon dating. The result of this test confirmed the suggested date range 17th to 18th centuries.

Sample Ref.	BM Lab. Ref. No.	Uncalibrated date years BP)	Calibrated Possible* Date Range
Bos R/KJB	BM-2156R	290 ± 110	1460-1675 cal AD or 1750-1795 cal AD

* Calibration made using the curve of Stuiver and Pearson, 1986, Radiocarbon, v.28, p. 805-838, and a standard error term with endpoints rounded to 5 years (Dr S.G. Bowman, British Museum, Reserach Laboratory, pers. comm. 1988).

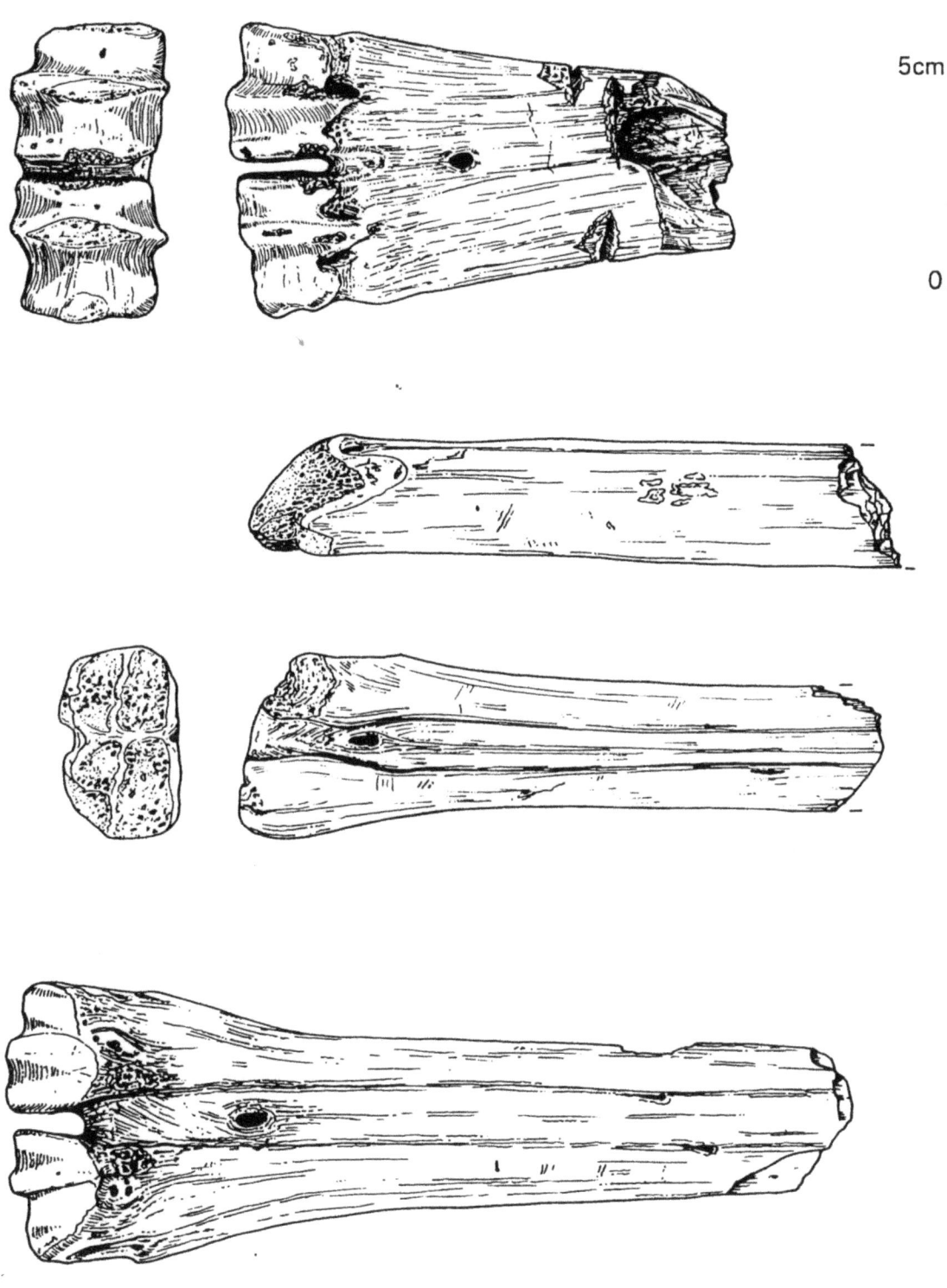

Figure 1. Cattle metapodial bones from "knuckle-bone floor", King John's House, Romsey, Hampshire, showing abrasion on the exposed part of the bones, the distal articular end.   Drawing: K.H. Armitage.

Present status: Extant; although much of the floor is covered by modern wooden floorboards, a small section has been left exposed and is illuminated for public viewing.

Sources of information: Apart from the very brief mention of this floor in the guidebook to "King John's House" (produced by the "King John's House Management Committee") this fine example of a surviving "knuckle-bone floor" is unpublished. Background historical information supplied by K. Stubbs, Historic Buildings Bureau, Hampshire County Council (1982, pers. comm.). Radiocarbon measurements carried out by the British Museum Research Laboratory (1983).

1.3 **Floor decoration in "follies"** - variation of the more traditional "knuckle-bone floors" found in private and public houses.

Site or place name: The "Bath House", Wrest Park, Silsoe, Bedfordshire.

Description: The floor is made up of smooth pebbles with an inlaid pattern formed by deer metapodial bones. All the bones are broken in half and have been embedded in the ground with their shaft end downwards and the articular end (proximal or distal) uppermost. The inlaid pattern takes the form of a central circle of proximal metacarpal and metatarsal bones, from which eight 'spokes' radiate outwards reaching into the corners of the room, which is built to a hexagonal plan. The bones forming the 'spokes' are the distal ends of metapodial bones. The folly was built in c. 1763. Formerly the design of the folly was attributed to Sir William Chambers, but it is now thought to be work of Stephens, one of his pupils (Kernow, pers. comm.).

Present status: The floor is open to public view (the estate has been designated as a national monument under the guardianship of the D.O.E., and is leased by them to the National Institute of Agricultural Engineering).

The floor was repaired some 20 years ago. During the restoration work, approximately half the bones present were found to be damaged and so were replaced with modern deer metapodial bones obtained from Richmond Park. In order to protect the surface of the floor from further damage, mortar was spread between the bones and pebbles to consolidate them. The modern bones are easily identifed by their fresh appearance. All the other bones are darker in colour but they still show no evidence of any real wear, suggesting that even these may not be the original 18th century bones.

Sources of information: Cited in Lysons (1806, Vol. 1). Historical and other background information supplied by A. Hunter, National Institute of Agricultural Engineering, Wrest Park (1981, pers. comm.) and P. Kernow, Principal Inspector, D.O.E. (1981, pers. comm.). Examined by P. Armitage (1982).

Site or place name: "Temple of Mercury", grounds of Moot House, Downton, Wiltshire.

Description: "Floor of cobbles divided into segments by bricks containing patterns formed by knuckle-bones" (Moore, 1982, pers. comm.).

The floor is thought to be contemporary with the "temple" (a 'folly') which was built in c. 1740.

Present status: Decorated floor and rest of folly is extant and is a local architectural curiousity.

Sources of information: N. Moore, Royal Commission on Historic Monuments (England), Salisbury, Wiltshire.

1.4 Patterned Floors in summerhouses and other minor "follies"

Site or place name: Summerhouse in the grounds of Darsham House, Saxmundham, Suffolk.

Description: Floor made up of hexagonal paving stones with the cheekteeth of horses forming a patterned effect in the diamond-shaped interstices. This feature is thought to date from the Regency period.

Present status: Floor still extant but wall and roof of the summerhouse were demolished some years ago.

Sources of information: Historical details supplied by Mr. R. Wetmore, owner of Darsham House. Floor inspected by P. Armitage (1984).

Site or place name: "Folly" in the grounds of Blaise Castle, near Bristol.

Description: Floor patterned with horse teeth.

Present status: A visit to the area in 1984 by P. Armitage and B. Levitan failed to find any evidence of this feature, which must have been demolished (date unknown).

Sources of information: B. Levitan, University Museum, Oxford (1984, pers. comm.).

2) CONSTRUCTIONAL WORKS - I Industrial Contexts

2.1 Horn core lined pits

Site or place name: Cutler's Gardens (TQ 3340/8150), City of London.

Description: Excavations carried out on this site between 1978 and 1980 by the Department of Urban Archaeology, Museum of London, revealed the floors of 17th century workshops (including a smithy) and other post-medieval debris of a wide variety of industries (bell founding, glass and clay tobacco pipe making, bone-working and ivory turning) sealed beneath East India Company warehouses (built at the end of the 18th century) whose shallow cellars ensured the survival of this evidence. In the northern part of the site, buried beneath the warehouses, was a series of at least thirteen pits whose function remains unknown. All of these pits, which are dated to the late 17th/early 18th century (not later than c. 1740), had their sides lined with cattle horn cores in distinct courses, with their tips all aligned one way (pointing outwards) and bonded with clay. By placing the inside curve of each core against the outside curve of its neighbour, the builder had managed to pack the cores forming each layer closely together without too much overlapping at their tips. Only at each corner of the pit were the cores arranges so that their tips interlocked by curving over each other - presumably such a configuration provided additional stability at this the weakest point in the internal lining to the pit.

The thirteen pits varied considerably in size with the largest of them measuring 8m long by 1.9m wide and 1.15m deep and the smallest 1.8m long by 0.8m wide and 1.14m deep. In every case the top of the pit was truncated by the contractors working on the site before the structure could be properly recorded by the archaeologists and so the true depths of the pits would have been greater than the values quoted above.

Five of the pits had traces of wooden planks (possibly pine) and oak joists projecting from the lining of horn cores, suggesting the possibility that either the structures were lined with wood, in the form of an enclosed wooden tank, or that they had originally been fitted with lids. All the pits were filled with the same dark silty material that on analysis was found to contain very little artefactual or biological material. Samples of the seeds examined by Straker and Davis (1982) were identified as common components of urban plant macrofossil assemblages. In addition to the seeds, the fills contained arthropods including wood-boring weevils and various beetles which live on dung or carrion, with a small number of phytophagus beetles hinting at the presence of vegetation used for bedding or stalling animals (Girling, 1982). Chemical analysis carred out by Evans (North East London Polytechnic) revealed that the pit fills were certainly not cess and that some of them contained traces of residue from bronze casting. Human, cattle and rabbit hair fibres, identified by Armitage, were also extracted from the fills. Despite the detailed investigation into the nature of the material filling these pits, their function remains unknown, but presumably they were used for some industrial purpose, for example as casting (founding) pits or as soakaways. The cattle hore cores used in the construction of these pits presumably came from the horner's workshops situated along Petticoat Lane (now Middlesex Street) and the butchers' shops in Aldgate High Street. The measurements of the horn cores are held by the author.

Present status: All these pits were destroyed in the course of redevelopment of the site. A sample of 539 horn cores was, however, collected and is held by the Museum of London.

Sources of information: Site details supplied by Ms Alison Balfour Lynn (site supervisor of the 1978 watching brief) and S. O'Connor Thompson (supervisor of the 1979-80 excavations), both of the Department of Urban Archaeology, Museum of London. Specialist reports provided by the following: Ms Vanessa Straker and Ms Anne Davis, DUA, Museum of London; Ms Maureen Girling, D.O.E. Ancient Monuments Laboratory; Dr. J. Evans, Chemistry Dept., North East London Polytechnic. Pits and associated horn cores inspected by P. Armitage. Brief description in Armitage, Davis, Straker & West (1983, 24-26).

Site or place name: Crosswall (TQ 3360/8101), City of London.

Description: During excavations carried out by the Department of Urban Archaeology, Museum of London, between September 1979 and March 1980 (site supervisor John Maloney), traces of two pits of 17th/18th century date were discovered. The sides of these pits had been lined with cattle horn cores laid neatly in course and separated by layers of clay. One pit had traces of wooden planking (soft wood) nailed to oak cross members (joists) along the top course of horn cores; possibly the remains of a lid or walkway.

Present status: Both pits were destroyed during redevelopment of the site. A sample of 28 horn cores was retained and is held in the collections of the Museum of London.

Sources of information: Pits and associated horn cores inspected by P. Armitage. Unpublished.

Site or place name: Mansell Street (TQ3388/8116), City of London.

Description: During excavations carried out by the Department of Urban Archaeology, Muesum of London in 1982 (site supervisor Anne Upsom) three pits of c. 1700 date were found; all three had been lined with cattle horn cores and were similar in construction to the pits recorded at Cutler's gardens.

Present status: All three pits destroyed during redevelopment of the site. A sample of horn cores was retained and is held in the collections of the Museum of London.

Sources of information: Pits and associated horn cores inspected by P. Armitage. Unpublished.

Site or place name: Gardener's Corner (TQ3381/8126), City of London.

Description: During excavations carried out by the Inner London Unit in 1980 two horn cores lined pits of 17th/early 18th century date were discovered. The function of these pits could not be determined, but they were probably connected with some industrial process.

Present status: Both pits were destroyed during redevelopment of the site. A sample of 40 horn cores was retained and is held by the Inner London Archaeology Unit.

Sources of information: Information on the site supplied by R. Whytehead, Director of the Inner London Unit (1980, pers. comm.). Pits and associated horn cores inspected by P. Armitage.

Site or place name: 6-7 Crescent (TQ 3361/8082) City of London.

Description: Excavations carried out by the Department of Urban Archaeology, Museum of London in April-July 1985 revealed two further horn core lined pits constructed and back-filled in the late 17th century. The pits were perhaps used for some unidentified industrial purpose.

Present status: Destroyed during redevelopment of the site.

Sources of information: Cited: a brief interim note on this site has been published by A. Westman (site supervisor) in Richardson (1986: 158).

3) CONSTRUCTIONAL WORKS - II Domestic and other non-industrial examples

3.1 **Packing/hardcore in house walls**

Site or place name: Privately owned cottage, 4 Little St. Mary's, Long Melford, Sudbury, Suffolk.

Description: During alterations to convert the cottage into a hairdressing salon in 1970, the owner, Mrs. H. Schofield, reported the discovery of animal bones embedded in the internal wall. These bones were later identifed by Dr Juliet Clutton-Brock and Dr Caroline Grigson as cattle metacarpal and metatarsal bones from adult and juvenile animals. These bones were complete and had not been butchered in any way (cf. metapodial bones used in the construction of "knuckle-bone floors").

According to Mrs H. Ross, Assistant Curator, Ipswich Museum and Art Galleries (1981, pers. comm.) the metapodial bones were embedded in the wall on either side of the doorway, four singly on either side and then in pairs, about 18 in (457 mm) apart, slightly staggered and parallel with the line of the wall. Mrs Ross is of the opinion that the bones served no structural purpose. If one accepts this observation, it follows that the bones had probably been placed in the wall to act as talismen protecting the house and its occupants from evil influences and general misfortune - a widespread practice in medieval and Tudor times. In my opinion, however, the presence within one wall of so many bones argues against this idea; instead of placing them in the wall as some form of "magic device" the builder probably simply used the bones as packing/hardcore material to bond and strengthen the construction.

Date: although the main building was constructed in the early 16th century, the wall in which the bones were found forms part of a later outshot which Mrs Coleman, an expert on Suffolk vernacular architecture, has dated to the 18th century (Ross, 1981, pers. comm.).

Present status: Part of the wall only was taken down during the alterations to the cottage; the remaining section survives with the bones left in situ. Samples of the bones extracted from the demolished section are held in the collections of the British Museum (Natural History) and Ipswich Museum.

Sources of information: Historical details supplied by Mrs H. Ross, Assistant Curator, Ipswich Museum and Art Galleries; information on the bones is based on the report of the study carried out by Dr Clutton-Brock and Dr Grigson, recorded in BM(NH) General Letter File (osteology room) 1970.7.

Site or place name: "Knuckle-bone house" (also known locally as "Whale bones stores", Cley, Norfolk.

Description: During refurbishment of the house a few years ago workmen discovered that animal bones (species and elements not specified) had been used as hardcore in the flint walls.

Present status: See 1.1 above.

Sources of information: Owners of the house, Mr and Mrs P. Dearden (1982, pers. comm.).

3.2 **In repair works** - where damage had occurred to the fabric of a building or part of a floor etc.

Site or place name: Privately owned cottage, formerly in Church Street, Ware, Hertfordshire.

Description: During excavations of the cellar of this 18th century cottage, the Ware Rescue Archaeology Group discovered a number of cattle metacarpal and metatarsal bones embedded in a section of the internal basement wall. It would appear that the flint wall in the west side of the cellar to the cottage had at one time suffered some structural damage (possibly during alterations to the building). This damage extended 1.8m along the top of the wall, to a depth of 0.2m; and had been repaired using cattle metapodial bones as replacements for the dislodged flints. The bones were laid lengthwise across the width of the wall, in distinct courses set in mortar, with their articular ends flush with the inner face of the wall (see p. 148).

The bones used in this repair work probably came from the nearby slaughterhouse which was contemporary with the cottage.

Present status: Destroyed in 1973 during redevelopment of the site.

Sources of information: Interim note (with photographs) published by Crossby (1974: 175). Archaeological/historical information supplied by F. Crossby, Ware Rescue Group (1973, pers. comm.); bones identified by P. Armitage (1973); additional site details obtained from C. Partridge, Director, Hart Archaeology Unit (1982, pers. comm.).

Site or place name: Cottage, formerly an ale-house known as "The Hart's Horn Inn", Ash, Surrey.

Description: During a visit to the cottage in 1926, the Rev. H. R. Huband reported finding horse teeth used in the repair of a section of the cellar floor:

"...in the cellar at the west end I found horses' teeth set upright in squares like tiling - a poor makeshift for the missing tiles" (Huband, 1926: 75).

The building was constructed in the 16th century, but the floor with the horse teeth probably dates from the 18th century.

Present status: Not known.

Sources of information: Huband (1926). P. Armitage is indebted to Mrs. Alison Locker for drawing his attention to this reference.

Site or place name: Great Blakenham Church, near Ipswich, Suffolk.

Description: Bones embedded in mortar in a stone wall to the churchyard. According to local village oral tradition, the wall contained human bones from the churchyard; it is more reasonable to suppose that these were animal bones obtained from a local butcher's slaughteryard for the purpose of repairing a damaged section in the wall. Some of the bones used to replace the dislodged flint nodules were "knuckle-bones". Date of repair work unknown.

Present status: No longer extant; the wall was demolished in 1968 in the course of a road widening scheme.

Sources of information: Unpublished; information obtained from the Rev. H.S. Excell, Rector of Great Blakenham (1981, pers. comm.) and Mrs H. Ross, Assistant Curator, Ipswich Museum and Art Galleries (1981, pers. comm.).

3.3 FASTENING SLATES TO A CHURCH ROOF

Site or place name: Parish Church, Pannal, near Harrogate, Yorkshire.

Description: Local parish account for Pannal, for the year 1678, records that "..Henry Winterburne was paid 17s for slating the church and 1s.3d. for 500 sheep shanks.." (Jackson, 1966). Here the term sheep shanks refers to sheep tibiae which were being used in lieu of wooden or metal pins to fasten the roofing slates.

Present status: Not known. Jackson (1966) remarks that he personally knew of no actual surviving examples of this practice in West Riding (see however 5.2 below).

Sources of information: Jackson (1966) who quotes from an earlier article by Hambletonian in The Meat Trades Journal for November 18th, 1965. P. Armitage is indebted to Dr M Ryder for drawing his attention to this example and for providing a copy of the note by Jackson (Ryder, 1984, pers. comm.).

4) CONSTRUCTIONAL WORKS - III Miscellaneous categories

4.1 **Cesspits and soakaways**

Site or place name: Greyfriars (site B), Oxford.

Description: During excavations carried out by the Oxford Archaeological Excavation Committee in 1969 (site supervisor J. Haslam) a sub-rectangular horn core lined pit was discovered. Clay tobacco pipes, pottery and glassware recovered from the fill to the pit suggest that this feature was constructed and back-filled between c. 1750-1820. Its original function was suggested by the organic nature of the fill which was probably cess.

Two tanneries are known from documentary sources to have operated near the Greyfriars site; the earlier one was active during the 17th and 18th centuries, and the later, in the late 18th and early 19th centuries. As the date for the pit based on the archaeological evidence spans the period c. 1750-1820, either one of these two tanyards is a candidate for the source of the horn cores used in its construction.

Present status: Pit destroyed in redevelopment of the site. A sample of 336 horn cores was collected for future reference, and these were presented to the British Museum (Natural History) (reg. nos. 1977.5014 to 5061). A preliminary interim (level III) report on these cores was prepared by P. Armitage (1982).

Sources of information: Site and historical information supplied by R. Wilson and M. Mellor, both of the Oxford Archaeological Unit. Horn cores examined by P. Armitage. Unpublished.

Site or place name: Church Street (TQ 391836), London Borough of West Ham.

Description: Excavations carried out by Passmore Edwards Museum in 1973 uncovered a 17th/18th century pit lined at the bottom with cattle metapodial bones. These bones were all complete (unbroken), and were aligned vertically with the proximal ends uppermost, forming a uniform surface. It is thought that such an arrangement enabled the bones to act as a filter, the whole pit serving as a soakaway for liquid waste (though whether this derived from industrial or domestic activity could not be determined from the scanty archaeological evidence).

It may be suggested that the cattle metapodial bones probably came from a nearby slaughteryard or tanyard (cf. the recently discovered post-medieval soakaway in Southwark where all of the cattle metapodial bones forming the bottom filter layer had been sawn in half - clearly indicating that they had derived from discarded bone-working waste).

Present status: Pit destroyed during redevelopment of the site. A sample of 35 metapodial bones (18 metcarpals and 17 metatarsals) was collected for future reference and these are held in the collections of the Passmore Edwards Museum.

Sources of information: Site information supplied by Ms Pat Wilkinson, Passmore Edwards Museum. Bones examined by Mrs. Alison Locker, Institute of Archaeology and P. Armitage (1984). Unpublished.

4.2 ROAD FOUNDATIONS

Site or place name: City of London and its suburbs (no specific localities recorded).

Description: During the 18th century cattle horn cores were spread out on the roads and then covered with sand and earth to form a "firm and durable" surface.

Present status: Not known. No examples of this practice have so far been discovered during archaeological excavations carried out in the City and Greater London.

Sources of information: Published eyewitness account of this unusual use of cattle horn cores by Kalm (1892: 73).

5) AGRICULTURAL AND HORTICULTURAL USES

5.1 **Agricultural land drains**

Site or place name: Forest House Estate, Epping Forest, Essex.

Description: Workmen digging drainage ditches on the estate in 1894 came across examples of late 17th century land drains lined with cattle horn cores (mostly from longhorn cattle). McKenny Hughes (1896) published a description of the drains based on details given him by his friend Mr Francis Barclay of Leyton, Essex: "... the horn cores were found in clayey soil about 14 in [0.35m] below the surface, arranged longitudinally three abreast, only one layer deep, and at the bottom of drains running down the slopes almost at right angles to the direction of the stream [at the bottom of the hill]".

Present status: Not known.

Sources of information: Published account by McKenny Hughes (1896).

Site or place name: 28-32 Upsdell Avenue (TQ 312917), Palmers Green, London Borough of Enfield.

Description: Accidentally discovered by C. Pratt in October 1978 when he was constructing a fish pond in his back garden. The site was subsequently investigated by the Enfield Archaeology Society, in 1978-80, who carried out excavations in Mr Pratt's back garden and in the two adjoining properties.

The excavations revealed the presence of two parallel rows of buried cattle horn cores, 10.8m apart, running east-west across the slope of the ground. These rows, interpreted as hollow land drains, were traced for a distance of 14m; the projected line continued across the hill and down the slope to where a brook once ran along the east side of the field. The horn cores had been laid into the clay subsoil at a depth below present day ground surface of 0.48m, with the tips embedded 0.08-0.10m into the clay. They had been placed at an angle of 45^0, with each core set up against the next so as to create a continuous row. Some of the cores were completely filled with black silt showing that the drain had become blocked and was no longer functioning. Date: Late 17th/early 18th century, based on pottery evidence and horn core typology.

Present status: Only about 10m of the two drains was destroyed during the course of the excavations; the remaining sections of this drainage system lie undisturbed under modern gardens and may be reinvestigated, if required.

Sources of information: Investigated by P. Armitage, R. Coxshall and J. Ivens (1978-1980); cited: Armitage, Coxshall & Ivens (1980).

Site or place name: Gray's Yard (TL (20) 233085), Batterdale, Old Hatfield, Hertfordshire.

Description: Examples of 17th/early 18th century horn core lined drains were discovered in 1968 by the Hatfield & District Archaeological Society. The horn cores were packed end to end in a double layer.

The drains were found close to the site of a former pond; it is suggested that the horn cores derived from a local tannery which used the water supply from the pond.

Delftware pottery found on the site suggests the drains may have been constructed and were in use some time during the last quarter of the 17th century and the first quarter of the 18th century. This date range is supported by the horn cores from the drains which match those from other late 17th/18th century contexts examined by P. Armitage.

Present status: Destroyed during redevelopment of the site.

Sources of information: Unpublished. Site details supplied by Ms Judith Harris and Ms Susan Fraser, both of Welwyn and Hatfield Museum Service (1983, pers. comm.).

Site or place name: Bethel Chapel, Hobbs Hill, Welwyn, Hertfordshire.

Description: When Bethel Chapel (erected in 1792) was being converted into a private house in 1959, there was "found under part of its floor a closely packed collection of animal horns [presumably cattle horn cores]..." These were thought to be the "remnants of a primitive drainage system" (Johnson, 1967: 39-40). This feature is probably best interpreted as an early agricultural land drain which occupied the site prior to the building of Bethel Chapel in the late 18th century - ie. the drain could perhaps be of 17th/early 18th century date.

Present status: Not known.

Sources of information: Johnson (1967).

Site or place name: Calcot Park Golf Club (SU 669725), Calcot Park, near Reading, Berkshire.

Description: The section of an undated land drain was uncovered by workmen in 1970, when the golf club was extending its club house. The drain took the form of a trench 1m deep, cut into the heavy clay soil (Swanmore series) of which the lower half was packed with the unbroken long bones of horse and cattle. The projected line of the ditch ran across Calcot Park from NW to SE following the natural slope of the ground.

Present status: Undisturbed sections of this drain remain buried beneath the gold course.

Sources of information: Unpublished. Description supplied by H. Carter, Museum and Art Gallery, Reading (1982, pers. comm.).

5.2 Farmhouse roofs

Site or place name: Villages in Wharfedale and neighbouring dales, Yorkshire (exact localities not recorded).

Description: In a letter to Dr Michael Ryder, Mr A. Raistrick mentions the widespread practice in Wharfedale and the neighbouring dales of securing stone slate roofing tiles in place using sheep tibiae in lieu of wooden or metal pins (cf. 3.3 above). According to the description given by Mr Raistrick, "The bone was not required to pierce the slater's laths, but only to hook behind them, and the taper of the tibia enable the bone to be driven to a tight hold in the hole chipped in the stone slate".

Present status: Not known; but probably many examples still survive today awaiting proper survey and documentation. It is interesting to note that oral tradition among certain field archaeologists hold that similar examples of this practice may still be found in Suffolk villages (Ms Miranda Armor-Chelu, 1985, pers. comm.). P. Armitage was, however, unable to discover any supporting evidence for this notion and this needs to be explored further.

Sources of information: Letter dated Sept. 22nd, 1966 from Mr A. Raistrick of Skipton, Yorkshire to Dr M. Ryder of the A.R.C. Animal Breeding Research Organisation, Roslin, Midlothian. P. Armitage is indebted to Dr Ryder for bringing this information to his attention and for providing a copy of the letter (Ryder, 1984, pers. comm.). Cited: a brief mention of the use of sheep tibiae as roof-tile pegs in the north of England appeared in Ryder (1984: 282).

Site or place name: Farmhouses in north east County Antrim, Northern Ireland (exact localities not recorded).

Description: Bone pegs were inserted into the walls of farmhouses, 6in (152mm) below the level of the eaves; these served as anchoring points for thatching ropes.

Present status: Old local practice, no longer followed. Relic examples may still be found today in the walls of derelict (ruined) farmhouses in this part of Northern Ireland (Gailey 1975, pers. comm.).

Sources of information: Description and historical details supplied by A. Gailey, Society for Folk Life Studies, Ulster Folk and Transport Museum (1975, pers. comm.).

5.3 Walls built around meadows and market gardens

Site or place name: Lane leading from St Albans town centre to Sheffield Mill, St Albans, Hertfordshire.

Description: Mundy (1904: 159) recorded that "...In a dark narrow lane leading from St Albans to the back meads watered by the river Vernon (sic) the way to Sheffield Mill is to be observed, although almost concealed by the obtrusion of ivy and other parasitical plants, a curious old wall, which on closer examination, proves to be composed wholly of the asseous (sic) remains of the horns of cattle. This singular structure has the appearance of being of very great antiquity, but no person living in the neighbourhood can give any correct account of its origin. Rumour asserts that some centuries ago a tanner resided near the spot, who purchased a plot of meadow land contiguous to his factory to build upon, and that, either in a spirit of eccentricity or from penurious motives, with a view to avoid the expense of bricks etc..., caused the wall in question to be erected from an accumulation of horns which he had lying by him in his tanning yard for many years".

No date is given for this feature, but it is documented that a large tannery was once located in nearby Fishpool Street, in 1744 (see Saunders, 1977: 11).

Present status: No longer extant.

Sources of information: Mundy (1904: 159) who quotes an article that appeared in an old Hertfordshire newspaper (not named) between 1840 and 1850.

Site or place name: Outer suburbs of London (exact locations not recorded).

Description: Earth walls incorporating cattle horn cores, built around meadowns and market gardens.

Kalm (1892: 62-70), a Swedish visitor to Britain, provides an eyewitness account of the mode of construction of these features: "An earth wall is cast up in the usual way. The breadth or thickness at the ground is made proportionate to the height of the intended face, for the higher the wall the broader the basis. When the earth has been cast up to a height of about six inches (152mm) it is levelled all over the top. Thereupon they have ready to hand a multitude of the quicks or inner parts of Ox-horns...This quick is so cut off that part of the skull commonly goes with it. The quicks are then set close beside one another over the earth that has been cast up for the wall, and this so that the larger and thicker ends of the quick, or that to which a portion of the skull is attached, is turned outwards or lies just in the face of the side of the wall. In this way two rows of quicks are laid, viz. one row on one side of the wall, and the other on the other, so that the small ends of the horn quicks meet in the middle. Over this afterwards cast earth about six inches (152mm) thick, when again in the forenamed manner is laid a stratum of double-ranged ox-horn quicks till the wall has reached the desired height (often to the height of six courses of horn cores)....The object of using these quicks is principally to bind the earth in the wall by them, and make it steady that it may not so soon slip down...Thus they know here to make use of that which in other places is thrown away".

Kalm (ibid.: 70) also mentions walls constructed of "bare ox-horn quicks .. piled up on one another as thick as ever they could find room, and the interstices only were filled up with mould".

Present status: No examples of any of these horn core walls have so far been discovered during archaeological excavations around the City of London; but the same technique of construction, based on lying courses of cattle horn cores bonded by clay, is known to have been adopted in the eastern parishes of London for strengthening the sides of industrial pits (see 2.1 above).

Sources of information: Published description in Kalm (1892: 69-70).

Site or place name: Private garden owned by Mr Peter Collinson, Peckham, London.

Description: Small fences around flower beds formed from horse and ox metapodial bones.

Kalm (1892: 67) gives an eyewitness account of this practice: "For the border or the outer edge of the flower-beds, Mr Collinson had set knuckle-bones, or horse or ox-legs...the transversal end was set down in the ground and the round curled end stood upwards. All were the same length, and quite close to one another, which performed the same service in hindering the earth from slipping down from the beds, as if there had been hoards set round them".

Present status: No examples of this type of bizarre flowerbed fence survive today.

Sources of information: Published account by Kalm (1892).

6) "ACOUSTIC RESONANCE VESSELS"

6.1 Horse skulls placed beneath floors

Site or place name: "Bungay House", Earsham Street, Bungay, Suffolk.

Description: During repair work to the house in the 1930s, the contractors discovered horse skulls lying under the floor boards. Ethel Mann visited the house in 1933 in order to investigate this discovery; she reported that on lifting two of the floorboards there "...beneath the (oak) joists were rows of horses' skulls, laid with great regularity, the incisor teeth of each resting on a square of oak or stone... The (floor) boards, which were of red pine, rested immediately upon the skulls". Mann estimated that the room contained up to forty horse skulls "..carefully prepared and boiled, and ...placed in position with great care and accuracy". Enquiries made by Mann revealed that in the house adjoining (once also part of "Bungay House") some twenty years previously workmen had also found horse skulls under the floor but that these had been removed.

Although "Bungay House" itself dated back to about 1620, Mann was unable to ascertain whether the horse skulls were contemporary or later additions to the building of the house. These skulls were thought by Mann to have given resonance to the floor when the room was being used by dancers (see example below).

Present status: Not known.

Sources of information: Mann (1934: 253-255) which includes a photograph of the horse skulls in situ.

Site or place name: Private house (formerly an inn) in Herefordshire (exact locality not recorded).

Description: Another example cited by Mann (1934) of horse skulls placed beneath a floor. Here the skulls were laid between the joists of the one room in the house with a boarded floor; all the other rooms had floors of flagstones. The object of burying the skulls under the floor was said to make "..a hollow sound when the dancers stamped their feet, as was the custom in some old country dances; the room in this house being known as 'the dancing room'".

Date of construction of this floor is not recorded.

Present status: Not known.

Sources of information: Published account by Mann (1934: 254).

Site or place name: Old inn called the "Portway" located eight miles west of Hereford.

Description: During a visit to the inn in 1852, Mr Thomas Blashill was told by the landlord that under the floor where they were standing were "two cartloads of horses' skulls" laid "to make the fiddle go better" - this being "a place where music and dancing sometimes went on". Returning to the same inn some twenty-eight years later, Mr Blashill discovered that some of these horse skulls had been unearthed by workmen during refurbishment of the building: "..twenty-four skulls were found screwed through the eye-holes to the underside of the floor boards in three rows" (see McKenny Hughes, 1915: 71).

Date of feature not indicated in Blashill's account.

Present status: Not known.

Souces of information: Account by T. Blashill originally published in Trans. R. Inst. Brit. Architects 1882, p. 83; quoted in McKenny Hughes (1915). P. Armitage is indebted to Tony Legge for bringing this example to his attention.

Site or place name: Thrimby Hall, Westmorland.

Description: In the year 1860 a quantity of horses' skulls was discovered under the parlour floor "where they had been placed for purposes of sound by the tenants who were a musical family".

Date of interment of the horses' skulls was apparently unknown, but local tradition held that the skulls derived from horses' heads "collected after a skirmish at Clifton Moor" - which perhaps suggests a Civil War (ie. mid 17th century) date (?).

Present status: Not known.

Sources of information: McKenny Hughes: Proc. Cambridge Antiquarian Society, Vol. XIX (no. LXVII, 1915: 70-71).

REFERENCES

ARMITAGE, P.L., COXSHALL, R. & IVENS, J. (1980) "Early agricultural land drains in the former parishes of Edmonton and Enfield". The London Archaeologist, vol. 3 (No. 15): 408-415.

ARMITAGE, P., DAVIS, A., STRAKER, V. & WEST, B. (1983) "Bugs, bones and botany Part 2". Popular Archaeology, Vol. 4 (No. 10): 24-27.

CHAMBERS, R. (ED) (1863) The Book of Days: A Miscellany of Popular Antiquities. 2 vols.

CROSSBY, F. (1974) "Ware". Hertfordshire Archaeological Review, No. 9: 175.

DURHAM, B. (1984) "The Thames crossing at Oxford". Oxoniensia, 49.

EVANS, G.E. (1966) The Pattern Under the Plough: Aspects of the Folk-Life of East Anglia. London: Faber & Faber.

GIRLING, M.A. (1982) "The arthropod assemblage from Cutler's Gardens, London". Ancient Monuments Laboratory Report 3670.

HUBAND, H.R. (1926) "The Hart's Horn Inn at Ash, Surrey". Surrey Archaeological Collections, XXXVII (Part 1): 75-78.

HURST, H. (1894) Proceedings of the Oxford Architectural and Historical Society Report for 1894 – Lent Term: 10-13.

INNOCENT C.F. (1916) The Development of English Building Construction Cambridge: University Press.

JACKSON, S. (1966) "Sheep shanks". Bradford City Art Gallery and Museums Archaeology Group Bulletin, vol.11 (No. 4): 1.

JOHNSON, W.B. (1967) Welwyn By and Large.

KALM, P. (1892) Visit to England (translated from the Swedish by J. Lucas from 1748 edn.).

LYSONS, D. (1806) Magna Britannia.

MANN, E. (1934) Old Bungay. London: Heath Cranton Ltd.

MCKENNY HUGHES, T. (1896) "On the more important breeds of cattle which have been recognised in the British Isles in successive periods". Archaeologia, 55: 125-158.

MUNDY, P. (1904) "A wall of horns at St. Albans". Home Counties Magazine, VI: 159.

RICHARDSON, B. (1986) "Excavation round-up 1985". The London Archaeologist, Vol. 5 (No. 6): 157-164.

RYDER, M.L. (1984) "Medieval animal products" Biologist, Vol. 31 (No. 5): 281-287.

SAUNDERS, C. (1977) "A sixteenth century tannery in St. Albans". Hertfordshire's Past, 3: 9-12.

SKELTON (1823) Oxonia Antiqua Restaurata.

SQUIRIS, T.W. (1928) In West Oxford: Historical Note and Pictures.

STRAKER, V. & DAVIS, A. (1982) "Seeds from the horn core lined pits at Cutler Street". Unpublished level III archival report, Museum of London.

www.ingramcontent.com/pod-product-compliance
Ingram Content Group UK Ltd.
Pitfield, Milton Keynes, MK11 3LW, UK
UKHW060200240426
12048UKWH00029B/1671

9 780860 545989